TUTORIAL CHE

9

Atomic
and Per

JACK BARRE

Imperial College of Sc. nce, Technology and Medicine, University of London

Cover images © Murray Robertson/visual elements 1998–99, taken from the 109 Visual Elements Periodic Table, available at www.chemsoc.org/viselements

ISBN 0-85404-657-7

A catalogue record for this book is available from the British Library

Published by The Royal Society of Chemistry, Thomas Graham House, Science Park, Milton Road, Cambridge CB4 0WF, UK
Registered Charity No. 207890
For further information see our web site at www.rsc.org

Typeset in Great Britain by Wyvern 21, Bristol
Printed and bound by Polestar Wheatons Ltd, Exeter

Preface

This book deals with the fundamental basis of the modern periodic classification of the elements and includes a discussion of the periodicities of some atomic properties and the nature of the fluorides and oxides of the elements. An introductory chapter deals with the chemically important fundamental particles, the nature of electromagnetic radiation and the restrictions on our knowledge of atomic particles imposed by Heisenberg's uncertainty principle. Atomic orbitals are described with the minimum of mathematics, and then used to describe the electronic configurations of the elements and the construction of the Periodic Table. A chapter is devoted to the periodicities of the ionization energies, electron attachment energies, sizes and electronegativity coefficients of the elements. There is also a section on relativistic effects on atomic properties. A brief overview of chemical bonding is included as the basis of the remaining chapters, which describe the nature and stoichiometries of the fluorides and oxides of the elements.

The book represents an attempt to present the periodicities of properties of the elements in a manner that is understandable from a knowledge of the electronic patterns on which the Periodic Table is based. It should be suitable for an introductory course on the subject and should give the reader a general idea of how the properties of atoms and some of their compounds vary across the periods and down the groups of the classification. This knowledge and understanding is essential for chemists who might very well find exceptions to the general rules described; such events being a great attraction in the continuing development of the subject. Apart from the underlying theoretical content, the general trends in periodicity of the elements may be appreciated by the simple statement that "size matters, and so does charge".

I thank Ellis Horwood for permission to use some material from my previous books, and Martyn Berry for helpful comments on the manuscript. I am very grateful to Pekka Pyykkö for his comments on the section about relativistic effects.

Jack Barrett
London

TUTORIAL CHEMISTRY TEXTS

This series of books consists of short, single-topic or modular texts, concentrating on the fundamental areas of chemistry taught in undergraduate science courses. Each book provides a concise account of the basic principles underlying a given subject, embodying an independent-learning philosophy and including worked examples. The one topic, one book approach ensures that the series is adaptable to chemistry courses across a variety of institutions.

TITLES IN THE SERIES

Stereochemistry *D G Morris*
Reactions and Characterization of Solids *S E Dann*
Main Group Chemistry *W Henderson*
d- and f-Block Chemistry *C J Jones*
Structure and Bonding *J Barrett*
Functional Group Chemistry *J R Hanson*
Organotransition Metal Chemistry *A F Hill*
Heterocyclic Chemistry *M Sainsbury*
Atomic Structure and Periodicity *J Barrett*
Thermodynamics and Statistical Mechanics *J M Seddon and J D Gale*

FORTHCOMING TITLES

Basic Atomic and Molecular Spectroscopy
Aromatic Chemistry
Organic Synthetic Methods
Quantum Mechanics for Chemists
Mechanisms in Organic Reactions
Molecular Interactions
Reaction Kinetics
X-ray Crystallography
Lanthanide and Actinide Elements
Maths for Chemists
Bioinorganic Chemistry
Chemistry of Solid Surfaces
Biology for Chemists
Multi-element NMR

Further information about this series is available at www.chemsoc.org/tct

Orders and enquiries should be sent to:
Sales and Customer Care, Royal Society of Chemistry, Thomas Graham House, Science Park, Milton Road, Cambridge CB4 0WF, UK

Tel: +44 1223 432360; Fax: +44 1223 426017; Email: sales@rsc.org

Contents

1 Atomic Particles, Photons and the Quantization of Electron Energies; Heisenberg's Uncertainty Principle

As an introduction to the main topics of this book – atomic structure and the periodicity of atomic properties – the foundations of the subject, which lie in quantum mechanics, and the nature of atomic particles and electromagnetic radiation are described.

Aims

By the end of this chapter you should understand:

- Which fundamental particles are important in chemistry
- The nature of electromagnetic radiation
- The photoelectric effect
- Wave–particle duality
- The relationship of electromagnetic radiation to changes of energy in nuclei, atoms, molecules and metals: the Bohr frequency condition
- The main features of the emission spectrum of the hydrogen atom and the quantization of the energies permitted for electrons in atoms
- Heisenberg's Uncertainty Principle and the necessity for quantum mechanics in the study of atomic structure

1.1 Fundamental Particles

To describe adequately the chemical properties of atoms and molecules it is necessary only to consider three fundamental particles: protons and neutrons, which are contained by atomic nuclei, and electrons which surround the nuclei. **Protons** and **neutrons** are composite particles, each consisting of three **quarks**, and are therefore not fundamental particles in the true sense of that term. They may, however, be regarded as being

The positive electron or **positron** has a mass identical to the negative electron (negatron) but has an opposite charge. Positrons are emitted by some radioactive nuclei and perish when they meet a negative electron, the two particles disappearing completely to form two γ-ray photons. Such radiation is known as **annihilation radiation**: $2e^- \rightarrow 2h\nu$

fundamental particles for all chemical purposes. The physical properties of **electrons**, protons and neutrons are given in Table 1.1, together with data for the **hydrogen atom**.

Table 1.1 Properties of some fundamental particles and the hydrogen atom (e is the elementary unit of electronic charge = $1.60217733 \times 10^{-19}$ coulombs)

Particle	*Symbol*	*Mass*[a]*/10^{-31} kg*	*RAM*	*Charge*	*Spin*
Proton	p	16726.231	1.0072765	$+e$	½
Neutron	n	16749.286	1.0086649	zero	½
Electron	e	9.1093897	0.0005486	$-e$	½
H atom	H	16735.339	1.007825	zero	–

[a] The atomic unit of mass is given by $m_u = m(^{12}C)/12$ kg. The relative atomic mass (RAM) of an element is given by $m_{element}/m_u$, where $m_{element}$ is the mass of one atom of the element in kg.

The absolute masses are given, together with their values on the **relative atomic mass (RAM)** scale, which is based on the unit of mass being equal to that of one twelfth of the mass of the ^{12}C isotope, *i.e.* the RAM of ^{12}C = 12.0000 exactly. The spin values of the particles are important in determining the behaviour of nuclei in compounds when subjected to **nuclear magnetic resonance spectroscopy** (NMR) and **electron spin resonance spectroscopy** (ESR).

Box 1.1 Representation of Atoms and their Nuclear Properties

The conventional way of representing an atom of an element and its nuclear properties is by placing the **mass number**, A [the whole number closest to the accurate relative mass (equal to the sum of the numbers of protons and neutrons in the nucleus)], as a left-hand superscript to the element symbol, with the nuclear charge, Ze^+, expressed as the number of protons in the nucleus (the **atomic number**, Z) placed as a left-hand subscript, $^{A}_{Z}X$, where X is the chemical symbol for the element. The value of $A - Z$ is the number of neutrons in the nucleus.

Mass numbers are very close to being whole numbers because the relative masses of nuclei are composed of numbers of protons and neutrons whose relative masses are very close to 1 on the RAM scale (see Table 1.1). The actual mass of an atom, M, can be expressed by the equation:

$$M = Zm_H + (A - Z)m_n - E_B/c^2 \quad (1.1)$$

where m_H is the mass of the hydrogen atom, and takes into account the protons and electrons in the neutral atom, and m_n is the mass of a neutron. E_B represents the **nuclear binding energy**, *i.e.* the energy released when the atom is formed from the appropriate numbers of hydrogen atoms and neutrons. This energy is converted into a mass by **Einstein's equation** ($E = mc^2$). For practical purposes the last term in the equation may be ignored, as it makes very little difference to the determination of the mass number of an atom.

For example, working in molar terms, the accurate relative atomic mass of the fluorine atom is 18.9984032 and the mass number is 19. The mass numbers of protons and neutrons are 1 and the 9 protons and 10 neutrons of the fluorine nucleus give a total of 19, the electron masses being too small to make any difference. Using accurate figures for the terms in equation (1.1), *i.e.* those from Table 1.1:

9 hydrogen atoms @	1.007825	9.070425
10 neutrons @	1.0086649	10.086649
	Sum	19.157074
	Accurate mass	18.9984032
	Difference	0.1586708
	Mass number	19

The possibility that a particular element with a value of Z may have varying values of $A - Z$, the number of neutrons, gives rise to the phenomenon of **isotopy**. Atoms having the same value of Z but different values of $A - Z$ (*i.e.* the number of neutrons) are **isotopes** of that atom. Isotopes of any particular element have exactly the same chemical properties, but their physical properties vary slightly because they are dependent upon the atomic mass. A minority of elements, such as the fluorine atom, are mono-isotopic in that their nuclei are unique. Examples of isotopy are shown in Table 1.2.

There is 0.015% of deuterium present in naturally occurring hydrogen, so the RAM value of the natural element (the weighted mean value of the RAM values for the constituent isotopes) is 1.00794.

Table 1.2 Examples of isotopy

Isotope[a]	*Z*	*A − Z*	*M*
^{1}H	1	0	1.007825
^{2}H (D)	1	1	2.014
^{3}H (T)	1	2	3.01605
^{35}Cl	17	18	34.968852
^{37}Cl	17	20	36.965903

[a] D = deuterium; T = tritium (radioactive, $t_{1/2}$ = 12.3 y by β^- decay to give ^{3}He).

Worked Problem 1.1

Q Use the data from Table 1.2 to calculate the RAM value for naturally occurring chlorine. 75.77% is ^{35}Cl.

A The RAM value for naturally occurring chlorine is 35.4527 (rounded off to four decimal places) because there is 75.77% of the lighter isotope in the natural mixture. The weighted mean mass of the two chlorine isotopes is given as:

$$\begin{aligned} \text{RAM(Cl)} &= 0.7577 \times 34.968852 + (1 - 0.7577) \times 36.965903 \\ &= 35.4527 \end{aligned}$$

The three chemically important fundamental particles are used to construct atoms, molecules and infinite arrays which include a variety of crystalline substances and metals. The interactions between constructional units are summarized in Table 1.3.

Table 1.3 A summary of material construction

Units	*Cohesive force*	*Products*
Protons, neutrons	Nuclear	Nuclei
Nuclei, electrons	Atomic	Atoms
Atoms	Valence	Molecules, infinite arrays
Molecules	Intermolecular	Liquid and solid aggregations

1.2 Electromagnetic Radiation

Electromagnetic radiation consists of "particles" or "packets" of energy known as **photons**. Evidence for this statement is described later in this chapter. The photon energies vary so considerably that the **electromagnetic spectrum** is divided up into convenient regions which are connected with their effects on matter with which they might interact. The regions range from **γ-rays**, through **X-rays**, **ultraviolet radiation**, **visible light**, **infrared radiation**, **microwave radiation** to long wavelength **radio waves**. The various regions are distinguished by their different **wavelength**, λ, and **frequency**, ν, ranges as given in Table 1.4. This table also includes the types of events that are initiated when the particular radiation interacts with matter.

Table 1.4 The regions of the electromagnetic spectrum. Wavelengths are given in metres, frequencies in hertz; 1 Hz = 1 cycle s^{-1}

Region	*Wavelength, λ/m*	*Frequency, ν/Hz*	*Effect on matter*
Radio	>1.0×10^4 to 1.0	1.0 <3.0×10^4 to 3.0×10^8	Changes in nuclear and electronic spins
Microwave	1.0 to 1.0×10^{-3}	3.0×10^8 to 3.0×10^{11}	Heating and changes in rotational energy of molecules
Infrared (IR)	1.0×10^{-3} to 7.0×10^{-7}	3.0×10^{11} to 4.3×10^{14}	Heating and changes in vibrational energy of molecules
Visible	7.0×10^{-7} to 4.0×10^{-7}	4.3×10^{14} to 7.5×10^{14}	Changes in electronic energy
Ultraviolet (UV)	4.0×10^{-7} to 1.0×10^{-9}	7.5×10^{14} to 3.0×10^{17}	Changes in electronic energy
X-ray	1.0×10^{-9} to 1.0×10^{-10}	3.0×10^{17} to 3.0×10^{18}	Ionization
γ-ray	1.0×10^{-10} to <1.0×10^{-10}	3.0×10^{18} to >3.0×10^{18}	Ionization

Table 1.5 Wavelength ranges of visible light

Colour	*Wavelength range/nm*
Red	720–640
Orange	640–590
Yellow	590–530
Green	530–490
Blue	490–420
Violet	420–400

[a] nanometre (nm) = 1×10^{-9} m.

The **visible region** is divided up into sections associated with colours which shade imperceptibly into one another. The colours, together with their wavelength ranges, are given in Table 1.5. The limits are to some extent dependent upon the characteristics of the eyes of the observer.

The boundaries between the various regions are somewhat arbitrary, except for the visible region which is self-evident, although this differs slightly from person to person.

The difference between the X and γ regions is one of scale and the methods by which the photons are produced. X-rays are emitted by some radioactive elements, but are mainly produced by directing a beam of energetic (*i.e.* accelerated) electrons on to a metal surface (*e.g.* copper). γ-Rays are generally emitted as the result of secondary processes following the primary decay of a radioactive element, *i.e.* the primary

X-rays used to be called Röntgen rays after their discoverer, who was awarded the 1901 Nobel Prize for Physics, the first year the prizes were awarded.

emission of either **α particles** (helium-4 nuclei) or **β particles** (usually negative electrons, although positron emission occurs in the decay of many neutron-deficient synthetic radioisotopes). The nuclear rearrangement of the **nucleons** (protons and neutrons) after the initial emission of the alpha or beta particle allows the nuclear ground state of the daughter element (*i.e.* the one produced by the decay of the parent nucleus, parents always having daughters in radiochemistry!) to be achieved by the release of the stabilization energy in the form of a γ-ray photon.

Box 1.2 The Relationship between Frequency and Wavelength

The relationship between frequency and wavelength is given by the equation:

$$\nu = c_0/\lambda_0 \tag{1.2}$$

where c_0 and λ_0 represent the speed of light (c_0 = 299792458 m s^{-1}) and its wavelength, respectively, in a vacuum. The speed of light, c, and its wavelength, λ, in any medium are dependent upon the refractive index, n, of that medium according to the equations:

$$c = c_0/n \tag{1.3}$$

and:

$$\lambda = \lambda_0/n \tag{1.4}$$

Under normal atmospheric conditions, n has a value which differs very little from 1.0, *e.g.* at 550 nm in dry air the value of n is 1.0002771, so the effect of measuring wavelengths in air may be neglected for most purposes. The frequency of any particular radiation is independent of the medium. Quoted frequencies are usually either numbers which are too small or too large when expressed in hertz, and it is common for them to be expressed as **wavenumbers**, which are the frequencies divided by the speed of light. This is expressed by the equation:

$$\tilde{\nu} = \frac{\nu}{c_0} \tag{1.5}$$

Wavenumbers, represented by the symbol, $\tilde{\nu}$ (nu-bar), have the units of reciprocal length and are usually quoted in cm^{-1}, although the strictly S.I. unit is the reciprocal metre, m^{-1}. For example, 3650 cm^{-1} = 365000 m^{-1}. A wavenumber represents the number of full waves that occur in the unit of length. For example, the wavenumber for the stretching vibration of a C–H bond in a hydrocarbon is around 3650 cm^{-1}, which means that there are 3650 wavelengths in one centimetre (see Figure 1.1).

In later equations the subscript is dropped from the symbol for the speed of light in a vacuum because the effect of the usual medium, the atmosphere at normal pressure, is very slight.

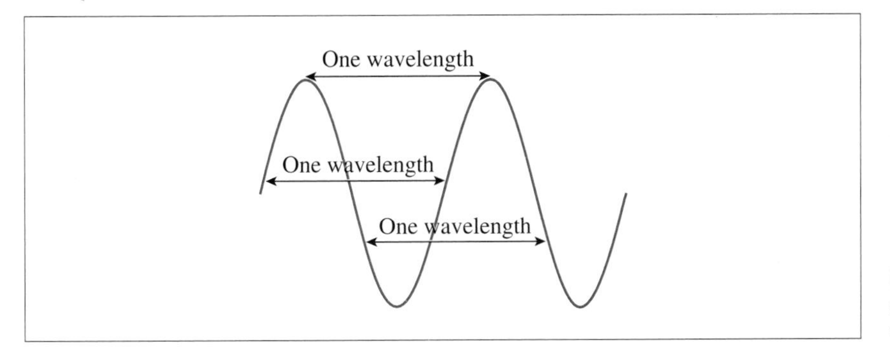

Figure 1.1 A representation of a wave motion, showing the wavelength, λ

Electromagnetic radiation is a form of energy which is *particulate* in nature. The terms wavelength and frequency do not imply that radiation is to be regarded as a wave motion in a real physical sense. They refer to the form of the mathematical functions which are used to describe the behaviour of radiation. The fundamental nature of electromagnetic radiation is embodied in quantum theory, which explains all the properties of radiation in terms of **quanta** or **photons**: packets of energy. A cavity, *e.g.* an oven, emits a broad spectrum of radiation which is independent of the material from which the cavity is constructed, but is entirely dependent upon the temperature of the cavity. Planck (1918 Nobel Prize for Physics) was able to explain the frequency distribution of broad-spectrum cavity radiation only by postulating that the radiation consisted of quanta with energies given by the equation:

$$E = h\nu \tag{1.6}$$

where h is Planck's constant ($6.6260755 \times 10^{-34}$ J s), the equation being known as the **Planck equation**. Radiation emitted from the liquid and solid states of matter is likely to be of the cavity type, but gases emit radiation which is characteristic of their individual nature as they consist of discrete molecules. In condensed phases (*i.e.* liquid or solid) the molecules or atoms are in constant contact so that the continual and varied perturbation of their energy levels allows the materials to act as Planck emitters. Such perturbation does not occur to any extent in the gaseous phase, so that any emission from gaseous species is that characteristic of the discrete molecules.

When dealing practically with equation (1.6) it is usual to work in molar quantities so that the energy has units of J mol^{-1}, and to achieve this it is necessary to use the equation:

$$E = N_A h\nu \quad (1.7)$$

where N_A is the Avogadro constant (6.0221367×10^{23} mol^{-1}).

Worked Problem 1.2

Q The wavelength of one of the lines emitted by a mercury vapour lamp is 253.7 nm. Calculate the quantum energy of this radiation.

A Using equation (1.2) to calculate the frequency of the radiation gives:
$\nu = c/\lambda = 299792458 \text{ m s}^{-1}/253.7 \times 10^{-9} \text{ m} = 1.18168 \times 10^{15} \text{ s}^{-1}$
Equation (1.7) gives the quantum energy as:

$E = N_A h\nu = 6.0221367 \times 10^{23} \text{ mol}^{-1} \times 6.6260755 \times 10^{-34} \text{ J s} \times 1.18168 \times 10^{15} \text{ s}^{-1} = 4.7153 \times 10^{5} \text{ J mol}^{-1} = 471.53 \text{ kJ mol}^{-1}$

Worked Problem 1.3

Q Derive an equation for converting the wavelengths of spectral lines into molar quantum energies.

A Combine equations (1.7) and (1.2) to give:

$$E = N_A h\nu;\ \nu = c/\lambda;\ \text{so } E = N_A hc/\lambda$$

1.3 The Photoelectric Effect

The quantum behaviour of radiation was demonstrated by Einstein (1921 Nobel Prize for Physics) in his explanation of the **photoelectric effect**. If radiation of sufficient energy strikes a clean metal surface, electrons (**photoelectrons**) are emitted, one electron per quantum. The energy of the photoelectron, E_{el}, is given by the difference between the energy of the incident quantum and the **work function**, W, which is the minimum energy required to cause the ionization of an electron from the metal surface:

$$E_{el} = N_A h\nu - W \quad (1.8)$$

the equation being expressed in molar quantities.

Photons with energies lower than the work function, *i.e.* when $N_A h\nu < W$, do not have the capacity to cause the release of photoelectrons. Equation (1.8) is based on the first law of thermodynamics (*i.e.* the law of conservation of energy). The photoelectric effect is shown diagrammatically in Figure 1.2, and Figure 1.3 is a representation of two cases of the possible interactions of photons with matter.

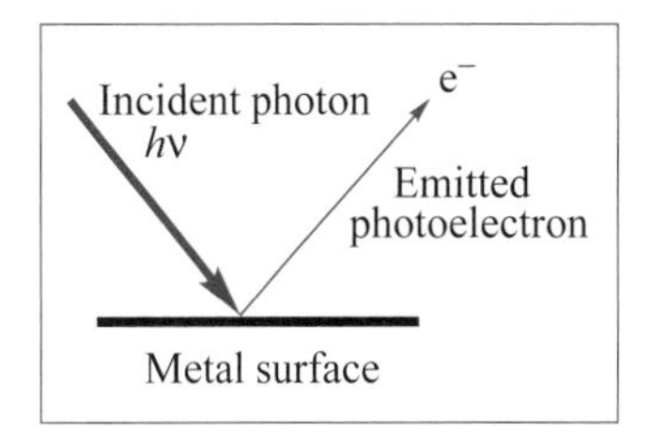

Figure 1.2 A diagrammatic representation of the photoelectric effect

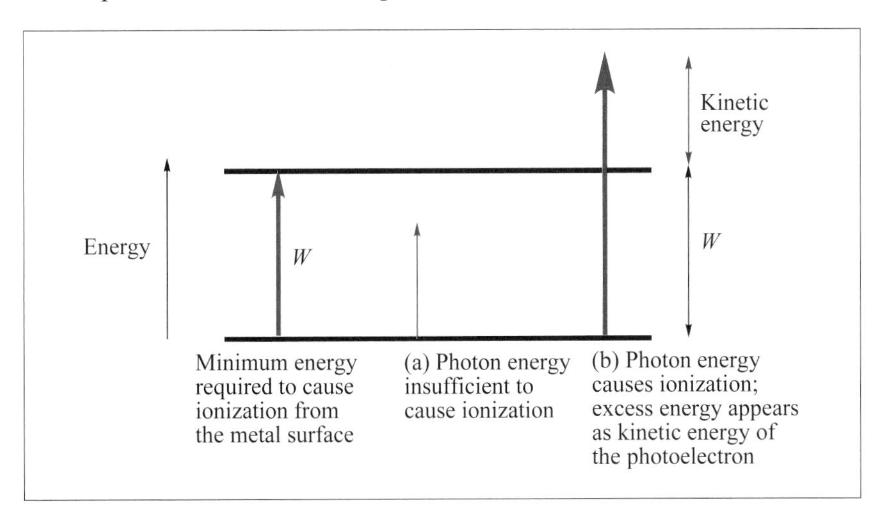

Figure 1.3 A representation of the interaction of two photons of different energies with a metal with a particular value of the work function. (a) The photon energy is not sufficiently large to cause electron emission; (b) the photon energy is large enough to cause photoelectron emission, the excess of energy appearing as the kinetic energy of the released electron

In Figure 1.3 the energy corresponding to the work function is represented by the energy level W with respect to the metal surface. In the case (a) of Figure 1.3 the photon energy, represented by the length of the arrow, is insufficient to cause the emission of an electron as the energy is less than that of the work function. In case (b) the photon has a large enough energy to cause the release of the photoelectron, and the excess of energy appears as the kinetic energy of the electron. Other important experimental results are that the number of electrons released in the photoelectric process is equal to the number of photons used, providing the photon energy is in excess of the work function and that the individual events are intimately related (one photon causing the immediate release of one photoelectron). An increase in the intensity of the radiation has no effect upon the kinetic energy of the photoelectrons, but it causes the number of electrons released per second to increase.

Photoelectrons can be released from gaseous molecules if radiation of appropriate energy is used, and measurements of their kinetic energies have provided much data about the energies of electrons in the molecules. Such energies are quantized, *i.e.* in any particular molecule the electrons are permitted to have only certain energies so that, for instance, if a molecule has three permitted levels of energy the molecule should have three different ionization energies in its **photoelectron spectrum**.

These experimental observations were in direct opposition to those expected for a **wave theory of radiation**. In wave theory, no threshold energy would be required for photoelectron release. A wave with low energy would simply operate long enough to contribute sufficient energy to cause the electron to be ionized. The kinetic energy of the photoelectrons would be expected to increase with the intensity of the radiation "waves".

When $N_A h\nu = W$ there is just enough photon energy to cause the release of a photoelectron; ν in this case is known as the threshold frequency

Another effect that the wave theory of radiation cannot explain is the

Early wave theorists proposed the existence of the "**ether**", which was the invisible (and non-existent) medium which would carry energy waves from the Sun to the Earth.

transmission of the Sun's rays through what is virtually a perfect vacuum between the star and the Earth in which there is nothing in which waves can form and carry the transmitted energy, unlike that which occurs in the oceans.

Worked Problem 1.4

Q Table 1.6 gives data for the kinetic energy of photoelectrons emitted by a calcium metal surface when irradiated by lines of the given wavelengths from a mercury lamp. Calculate the energy equivalents of the mercury lines used and, by plotting a graph of the kinetic energy of the photoelectrons versus the quantum energies, derive a value for the work function of calcium metal.

Table 1.6 Data for the photoelectric effect on a calcium metal surface

Wavelength of mercury line/nm	*Kinetic energy of photoelectrons/kJ mol^{-1}*
435.83	No electrons emitted
365.02	50.82
313.17	105.07
253.65	194.71
184.95	369.89

A The graph is shown in Figure 1.4 and indicates that the work function for calcium metal is 276.9 kJ mol^{-1}.

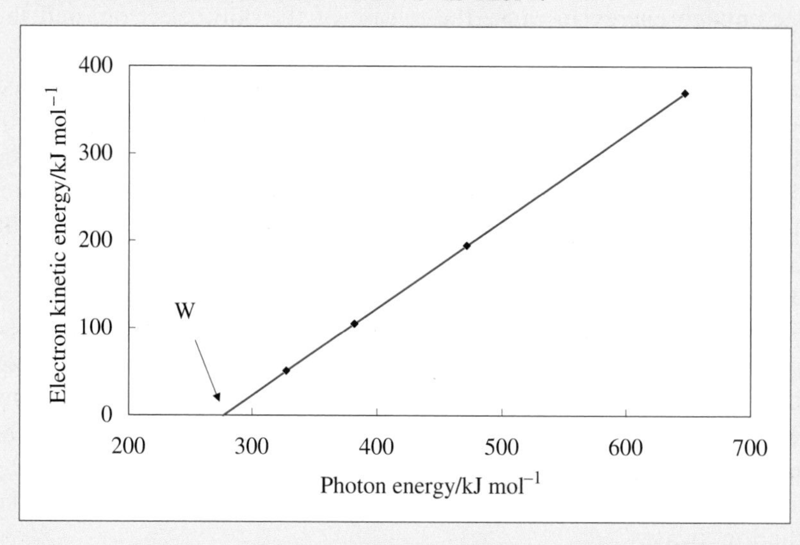

Figure 1.4 The results from a photoelectric effect experiment using a calcium metal surface irradiated by radiation of various lines given out by a mercury-vapour discharge lamp

1.4 Wave–Particle Duality

The photoelectric effect and the properties of cavity radiation show that the classical idea that electromagnetic radiation is a form of wave motion is defective. Interference and diffraction phenomena, in which electromagnetic radiation behaves as though the photons are governed by a wave motion, are understandable in the enhancement and enfeeblement of waves of probability of finding photons in particular localities. These phenomena are shown in Figure 1.5.

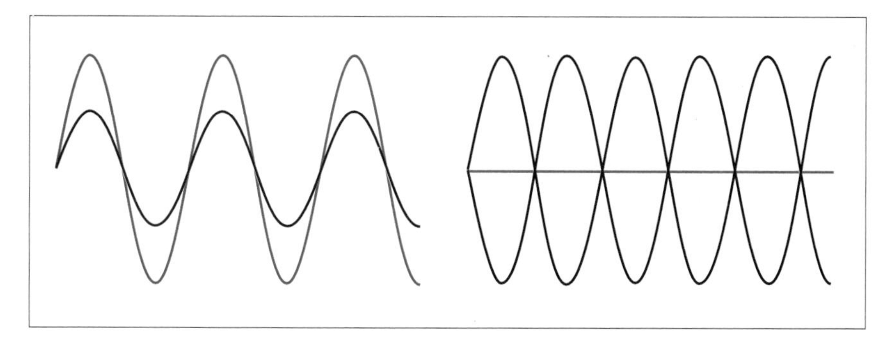

Figure 1.5 The enhancement (*left*) and enfeeblement (*right*) of waves which are in-phase and out-of-phase, respectively. In both cases the resultant of the waves' interactions are shown by the *red lines*

There is no reason to regard photons other than as "particles" or "packets" of energy, with particular properties that may be described in terms of mathematics that take the form of a wave motion. This does not imply that photons are waves.

Electrons are considered to be particles in the normal sense of the term in that they possess mass, and when incident upon a fluorescent screen produce a discrete scintillation or flash of light. Anyone who has ever watched a television programme has experienced this phenomenon. Electrons, however, and other atomic particles, can undergo the process of diffraction. Davisson and Germer first discovered electron diffraction in 1927 by allowing a beam of electrons to hit the surface of a crystal of nickel metal. A photographic plate was used as the detector for the diffracted electrons, and after electrons had struck it the plate showed a circular pattern of rings interspersed with regions which had not been affected.

If a beam of electrons with a particular energy, dependent upon the extent to which they have been accelerated in an electric field, is directed through a thin foil of metal (*e.g.* silver or gold), a photographic plate at the other side of the foil shows a circular diffraction pattern when developed. Experiments on metal foils were first carried out by G. P. Thompson and Reid in 1928, and Davisson and Thompson were awarded the 1937 Nobel Prize for Physics for their combined work.

That real particles could be diffracted and seemed to have wave properties interested Count Louis de Broglie. He solved the problem by proposing '**matter waves**' which, like electromagnetic waves, could interact by enhancement and enfeeblement and produce diffraction patterns. His logic depended upon the Planck equation (1.6) and Einstein's equation from his special theory of relativity relating energy and mass:

$$E = mc^2 \tag{1.9}$$

He then applied the two equations to a photon, equating the two right-hand sides to give:

$$h\nu = mc^2 \tag{1.10}$$

Rearrangement of equation (1.2), and dropping the subscript on c, gives the wavelength of a photon:

$$\lambda = c/\nu \tag{1.11}$$

If the frequency, ν, in equation (1.11) is replaced by the term mc^2/h derived from equation (1.10), the equation for the wavelength, λ, becomes:

$$\lambda = h/mc \tag{1.12}$$

This equation is theoretically exact for photons (regarded as "particles" of energy), but is not immediately applicable to atomic particles which cannot have a velocity equal to the speed of light. De Broglie's "quantum leap" in thinking, was to modify equation (1.12) by replacing the speed of light, c, with the speed of the atomic particle, υ, so that:

$$\lambda = h/m\upsilon \tag{1.13}$$

which may be written as:

$$\lambda = h/p \tag{1.14}$$

where p is the momentum, $m\upsilon$, of the particle. The equation faithfully reproduces the "wavelengths" of particles undergoing diffraction experiments.

De Broglie was awarded the 1929 Nobel Prize for Physics for his work on the wave nature of electrons.

The de Broglie equation (1.14) applies to the diffraction of other nuclear particles, but does not imply that the particles are "waves", just that their behaviour under diffraction conditions is governed by a mathematical probability wave-motion which allows for interactions of enhancement (waves in-phase) and enfeeblement (waves out-of-phase) to determine the diffraction patterns observed.

Worked Problem 1.5

Q Calculate the wavelength of an electron moving at 90% of the speed of light. The theory of relativity indicates that the mass of a moving object is dependent upon its velocity, according to Einstein's equation:

$$m = \frac{m_0}{\left(1 - \frac{\upsilon^2}{c^2}\right)^{1/2}}$$

where m is the mass of the particle moving at a velocity υ and m_0 is its rest mass.

A An electron moving at 90% of the velocity of light would have a mass:

$$m = 9.1013897 \times 10^{-31} \div \left(1 - \frac{0.9^2}{1}\right)^{1/2} = 2.0898373 \times 10^{-30} \text{ kg}$$

The wavelength of the electron would be:

$\lambda = h/m\upsilon = 6.6260755 \times 10^{-34}$ J s) $\div$ ($2.0898373 \times 10^{-30}$ kg $\times$ 299792458 m $s^{-1} \times 0.9$)
$= 1.175 \times 10^{-12}$ m = 1.175 pm

[1 J = 1 kg m^2 s^{-2}]

1.5 The Bohr Frequency Condition

The other major application of the Planck equation is in the interpretation of **transitions** between energy states or energy levels in nuclei, atoms, molecules and in infinite aggregations such as metals. If any two states i and j have energies E_i and E_j, respectively (with energy $E_i < E_j$), and the difference in energy between the two states is represented by ΔE_{ij}, the appropriate frequency of radiation which will cause the transition between them in absorption (*i.e.* if state i absorbs energy to become state j) or will be emitted (*i.e.* if state j releases some of its energy to become state i) is that given by the equation:

$$\Delta E_{ij} = E_j - E_i = h\nu \quad (1.15)$$

This modification of the Planck equation was suggested by Bohr (1922 Nobel Prize for Physics) and is sometimes referred to as the **Bohr frequency condition**. It applies to a large range of transitions between energy states of nuclei, electrons in atoms, rotational, vibrational and electronic changes in ions and molecules, and the transitions responsible for the optical properties of metals and semiconducting materials.

The interpretation of transitions in absorption and emission are shown in Figure 1.6.

The Planck equation (1.6) can be applied to the fixed (quantized) energies of two levels, E_j and E_i: $E_j = h\nu_j$ and $E_i = h\nu_i$, and the difference in energy between the two levels is:

$$\Delta E_{ij} = E_j - E_i = h\nu_j - h\nu_i = h\nu_{ij}$$

which is how equation (1.15) is derived.

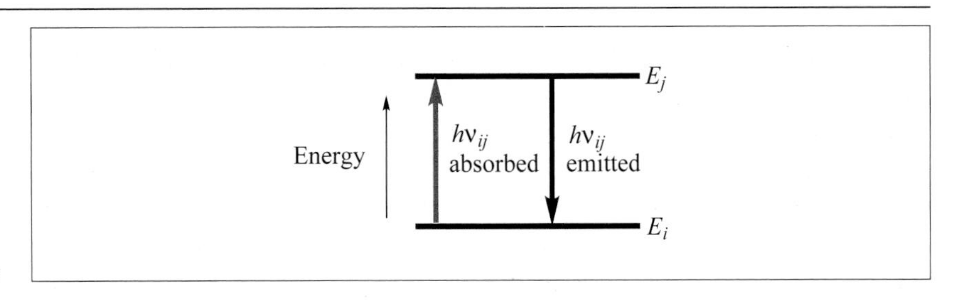

Figure 1.6 A representation of the Bohr frequency condition for the absorption and emission of radiation as an electron transfer occurs between two energy levels

1.6 The Hydrogen Atom

The hydrogen atom is the simplest atom, consisting of a single proton as its nucleus and a single orbital electron. Experimental evidence for the possible electronic arrangements in the hydrogen atom was provided by its emission spectrum. This consists of "**lines**" corresponding to particular frequencies rather than being the continuous emission of all possible frequencies. Discrete "line" emissions of characteristic wavelengths of light from electrical discharges through gases were observed in the early days of spectroscopy. The lines are sometimes regarded as monochromatic (*i.e.* of a precise wavelength), but they have very small finite widths that depend upon the temperature and pressure of the system.

Line spectra are so-called because of the way in which they are observed, the emitted radiation passing through a slit onto a photographic plate and registering as a "line".

The discrimination of emission frequencies leads to the concept of discrete **energy levels** within the atom that may be occupied by electrons. Detailed analysis of the wavelengths of the lines in the emission spectrum of the hydrogen atom led to the formulation of the empirical **Rydberg equation**:

$$\frac{1}{\lambda} = R\left[\frac{1}{n_i^2} - \frac{1}{n_j^2}\right] \tag{1.16}$$

where the wavelength, λ, relates to an electronic transition from the level j to the level i, j being larger than i, the terms n_i and n_j being particular values of what is now known as the **principal quantum number**, n. The relative energies of levels i and j are shown in Figure 1.6.

The Rydberg constant, R, has an experimentally observed value of 1.096776×10^7 m^{-1} for the hydrogen atom. Because it is the frequency, ν, of the radiation which is proportional to the energy, equation (1.16) can be transformed into one expressing the frequencies of lines by using the relationship given by equation (1.2):

$$\nu = \frac{c}{\lambda} = Rc\left[\frac{1}{n_i^2} - \frac{1}{n_j^2}\right] \tag{1.17}$$

The other term in the equation, c, is the velocity of light. The terms in brackets in equation (1.17) are dimensionless, and if R is multiplied by c (m s^{-1}) the result is a frequency (s^{-1} or hertz, Hz).

The line spectrum of the hydrogen atom is the basis of the concept of the quantization of electron energies, that is, the permitted energies for the electron in the atom of hydrogen are quantized. They have particular values, so that it is not possible for the electron to possess any other values for its energy than those given by the Rydberg equation. As that equation implies, the electron energy is dependent upon the particular value of the quantum number, n. The Rydberg equation can be converted into one which relates directly to electron energies by multiplying both sides of equation (1.17) by Planck's constant, h, and by the Avogadro constant, N_A, to obtain the energy in units of J mol^{-1}. Such a procedure makes use of the Planck equation (1.6), which relates the frequency of electromagnetic radiation, ν, to its energy ($E = h\nu$).

The Rydberg equation in molar energy units is:

$$E_{ij} = N_A h\nu = N_A Rch\left[\frac{1}{n_i^2} - \frac{1}{n_j^2}\right] \tag{1.18}$$

Bohr interpreted spectral lines in the hydrogen spectrum in terms of electronic transitions within the hydrogen atom. The Bohr equation (1.15) expresses the idea that $E_j - E_i$ represents the difference in energy between the two levels, ΔE, and may be written in the form:

$$E_{ij} = \Delta E = E_j - E_i = N_A h\nu_{ij} \tag{1.19}$$

A combination of equations (1.18) and (1.19) gives:

$$E_j - E_i = N_A Rch\left[\frac{1}{n_i^2} - \frac{1}{n_j^2}\right] \tag{1.20}$$

Equation (1.20) may be regarded as being the *difference* between the two equations:

$$E_j = -\frac{N_A Rch}{n_j^2} \tag{1.21}$$

and:

$$E_i = -\frac{N_A Rch}{n_i^2} \tag{1.22}$$

so that a general relationship may be written as:

$$E_n = -\frac{N_A Rch}{n^2} \tag{1.23}$$

Equation (1.23) describes the permitted quantized energy values for the electron in the atom of hydrogen. Some of these values are shown in Figure 1.7, together with the possible electronic transitions which form part of the emission spectrum of the atom. The reference zero for the diagram is the energy corresponding to the complete removal or ionization [to give the bare proton, $H^+(g)$] of the electron from the influence of the nucleus of the atom. The Lyman transitions are observed in the far-ultraviolet region of the electromagnetic spectrum (wavelengths below 200 nm). The Balmer transitions are found mainly in the visible region, and the Paschen transitions are in the infrared region. There are other series of transitions of lower energies which are all characterized by their different final values of n. The fundamental series is the Lyman series, in which all the lines have n_i values of 1. A summary of the n values for the first five series of lines in the hydrogen spectrum is given in Table 1.7.

Table 1.7 The n values for the first five series of hydrogen line spectra

Series	*Value of* n_i	*Values of* n_j
Lyman	1	2, 3, 4, ...
Balmer	2	3, 4, 5, ...
Paschen	3	4, 5, 6, ...
Brackett	4	5, 6, 7, ...
Pfund	5	6, 7, 8, ...

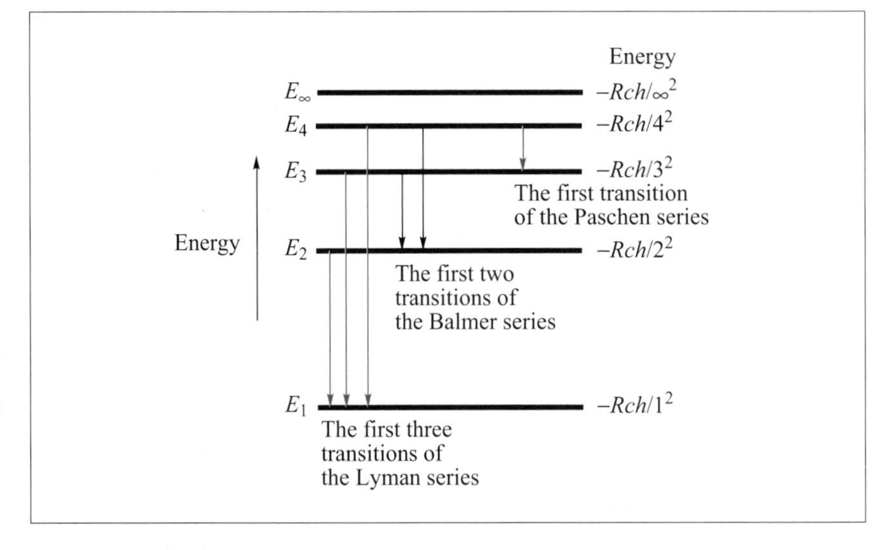

Figure 1.7 The lower four energy levels of the hydrogen atom and some of the transitions which are observed in the emission spectrum of the gaseous atom. The reference zero energy, corresponding to $n = \infty$, is indicated

Worked Problem 1.6

Q A line in the hydrogen atomic spectrum at a wavelength of 94.93 nm is a member of the Lyman series. Calculate the value of the principal quantum number of the energy level from which the spectral line is emitted.

A Algebraic manipulation of equation (1.16) is the best approach to the problem. Noting that the value of n_i is equal to 1 for the Lyman series, the equation may be written in the form:

$$\frac{1}{\lambda} = R\left[1 - \frac{1}{n_j^2}\right]$$

which may be rearranged to give:

$$n_j = \left(\frac{\lambda R}{\lambda R - 1}\right)^{1/2}$$

which gives n_j (the appropriate value of the principal quantum number) as 5 for the particular wavelength of the Lyman emission line.

1.7 The Observation of Electrons; the Heisenberg Uncertainty Principle

When electrons are observed by using a scintillation screen, *e.g.* one coated with zinc sulfide which scintillates (emits a localized flash of light when struck by an energetic particle such as an electron), they appear to be "particulate", since for each collision of a single electron a single flash of light is produced. If an electron beam is allowed to impact upon the surface of a metal crystal and a photographic plate detector is used, a diffraction pattern is produced, as though the electrons had "wave" properties. Similar dual properties are observed for photons. Light "waves" can be diffracted, but photons individually can cause photoionization of electrons from a metal surface. In the latter type of observations they seem to have "particulate" properties.

The behaviour of small atomic particles and of photons is described by the words we have available in the language, and both the words *wave* and *particle* have to be used carefully in this context. It would appear, therefore, that since various methods of observing small particles and photons lead to either wave or particulate views of their nature, the actual event of observation causes the variation of interpretation of the results, since one cannot imagine that small particles or photons change their properties according to the method used to observe them. In reality, part of the nature of small particles and photons is revealed by one method of observation and another part by another method, neither method revealing the whole nature of the particle being observed.

The majority of methods of observation involve photons which are used to illuminate the object. After bouncing off the object the photons are transmitted to a recording device, which may be a photographic plate, a photoelectric cell or the eye. Such methods are satisfactory for the majority of normal objects. The photons which are used do not in any measurable way affect the object. The situation is different with microscopic objects such as atoms and electrons. Electrons cannot be *seen* in the normal sense of that word. They are too small. To observe such small objects we should have to use a microscope with a resolving power greater than any actual microscope now in existence.

A useful exercise is to carry out a "thought experiment" in which an ideal (non-existent) microscope is used to view an electron. In order that the resolving power is suitable we have to use electromagnetic radiation that has a wavelength equal to or smaller than the object to be observed. To maximize the resolving power of the ideal microscope we may think of using short-wavelength γ-rays. These are photons with extremely large energies, and if they are used to strike the electron they will interact to alter the momentum of the electron to an indeterminate extent. In this experiment we may have observed the electron, but the process of "seeing" has altered the electron's momentum, and we come to the conclusion that if the position of the electron is known, the momentum is uncertain.

Next, consider the measurement of the velocity of the electron. To do this we have to observe the electron twice in timing its motion through a given distance. As we have already concluded, the process of "seeing" involves the transference of indeterminate amounts of energy to the electron and thus alters its momentum. To minimize this, we can use extremely long-wavelength, low-energy photons in the ideal microscope. This would ensure that the uncertainty in the momentum was minimal but, of course, with such long wavelengths the resolving power of the microscope is reduced to a minimum, and it would not be possible even with our "ideal" system to observe the position of the electron with any certainty. We conclude from this that if the momentum of the electron is known accurately, it is not possible to know its position with any certainty.

The two conclusions we have reached are summarized as, and follow from, Heisenberg's **Uncertainty Principle** (sometimes called **The Principle of Indeterminacy**) that may be stated in the form:

"It is impossible to determine simultaneously the position and momentum of an atomic particle."

Equation (1.24) is a simple mathematical expression of the uncertainty principle, in which Δp represents the uncertainty in momentum of the electron and Δq its uncertainty in position:

$$\Delta p \cdot \Delta q = h \tag{1.24}$$

The main consequence of the uncertainty principle is that, because electronic energy levels are known with considerable accuracy, the positions of electrons within atoms are not known at all accurately. This realization forces theoretical chemistry to develop methods of calculation of electronic positions in terms of probabilities rather than assigning to them, for example, fixed radii around the nucleus. The varied methods of these calculations are known collectively as **quantum mechanics**.

Summary of Key Points

1. The fundamental particles used in the construction of atoms were described.
2. Isotopy was explained.
3. The nature of electromagnetic radiation was described in terms of quanta or photons, and evidence for such a description is given from Planck's explanation of cavity radiation and the photoelectric effect.
4. The Bohr frequency condition was introduced, which relates the difference in energy between any two energy levels to the energy of a photon that is either absorbed or emitted in a radiative transition.
5. Wave–particle duality was discussed in terms of the wave-like and particulate properties of both electromagnetic radiation and electrons.
6. The Rydberg equation was described as evidence for discrete energy levels and as an example of the Bohr frequency condition. Using the Rydberg equation, a general equation for the electronic energy levels of the hydrogen atom was derived.
7. The difficulties of observing atomic particles were discussed, and this led to a statement of the Heisenberg uncertainty principle.

Problems

1.1. Calculate the RAM value for the element chromium, given that the natural element contains atoms with masses, correct to four decimal places, of 49.9461 (4.35%), 51.9405 (83.79%), 52.9407 (9.50%) and 53.9389 (2.36%).

1.2. Blue light has a wavelength of about 470 nm. Calculate (i) the energy of one photon, and (ii) the energy of one mole of photons (one mole of photons is termed an **Einstein**).

1.3. The work functions of the elements gold, vanadium, magnesium and barium are 492, 415, 353 and 261 kJ mol^{-1}, respectively. Calculate the threshold frequencies of radiation which will liberate photoelectrons from the surface of the elements.

1.4. Irradiation of the surface of a piece of clean potassium metal with radiation from a mercury vapour discharge lamp with wavelengths of 435.83, 365.02, 313.17, 253.65 and 184.95 nm causes the emission of photoelectrons with kinetic energies of 52.6, 105.8, 160.0, 249.7 and 424.9 kJ mol^{-1}, respectively. By plotting a graph of kinetic energy versus frequency, estimate a value for the threshold frequency and then calculate a value for the work function for the potassium metal surface. From the slope of the graph, calculate a value for Planck's constant.

1.5. The longest wavelength line of the Balmer series in the emission spectrum of the hydrogen atom is 656.3 nm. Use the Rydberg equation to calculate the wavelengths of (i) the second line of the Balmer series, (ii) the first line of the Paschen series and (iii) the first line of the Lyman series.

Further Reading

B. Hoffmann, *The Strange Story of the Quantum*, 2nd edn., Dover, New York, 1959. A highly readable book which gives insights to the concepts discussed in this chapter.

G. Herzberg, *Atomic Spectra and Atomic Structure*, 2nd edn., Dover, New York, 1944. This is a still-available classic text by the 1971 Nobel laureate for chemistry, and contains descriptions of atomic spectra and atomic structure which are "out of the horse's mouth!"

T. P. Softley, *Atomic Spectra*, Oxford University Press, Oxford, 1994. A very concise account of the subject that represents a suitable extension of the material in this chapter.

D. O. Hayward, *Quantum Mechanics for Chemists*, Royal Society of Chemistry, Cambridge, 2002. A companion volume in this series.

2
Atomic Orbitals

Atomic orbitals represent the locations of electrons in atoms, and are derived from quantum mechanical calculations. The calculations are only briefly outlined in this chapter, but the results are described in some detail because atomic orbitals are the basis of the understanding of atomic properties.

Aims

By the end of this chapter you should understand:

- That the energy levels of electrons in the hydrogen atom are quantized
- That the Schrödinger equation can be solved exactly for the hydrogen atom
- What is meant by an atomic orbital
- The quantum rules for describing atomic orbitals
- The spatial orientations of s, p, d and f atomic orbitals

2.1 The Hydrogen Atom

2.1.1 Energy Levels of the Electron in the Hydrogen Atom

The atomic spectrum of the hydrogen atom is described in Chapter 1. Its study, and that of other atomic spectra, provide much evidence for the **quantization** of electronic energy levels. The energy levels of the single electron in the hydrogen atom are represented by equation (2.1), which is derived from the Rydberg equation (1.16) in Chapter 1:

$$E_n = -\frac{N_A Rch}{n^2} \quad (2.1)$$

In addition, the Rydberg equation gives electronic energy levels with a high degree of accuracy. The actual permitted energies for the electron in a hydrogen atom are therefore known very accurately. The consequence of such accurate knowledge is that the position of the electron within the atom is very uncertain. As indicated by the uncertainty principle, a small uncertainty in momentum is related to a large uncertainty in position.

The consequence of being in considerable ignorance about the position of an electron in an atom is that calculations of the **probability of finding an electron** in a given position must be made. Other books in this series deal with the details of quantum mechanical calculations for atoms and molecules.

2.1.2 Quantum Mechanics and the Schrödinger Equation

The mathematical details of the setting up of the **Schrödinger equation** and its solutions are left for more specialist texts, and dealt with only briefly in this section.

The general mathematical expression of the problem may be written as one form of the Schrödinger wave equation:

$$H\psi = E\psi \quad (2.2)$$

The equation implies that if the operations represented by H (the **Hamiltonian operator**) are carried out on the function, ψ, the result will contain knowledge about ψ and its associated permitted energies. The term represented by ψ is the **wave function**, which is such that its square, ψ^2, is the **probability density**.

Strictly, this should be written as $\psi\psi^*$, where ψ^* is the complex conjugate of ψ, in case there is an imaginary component, *i.e.* expressions which contain "i", which is the square root of minus 1.

The value of $\psi^2 d\tau$ represents the *probability of finding the electron* in the volume element $d\tau$ (which may be visualized as the product of three elements of the Cartesian axes: $dx.dy.dz$). Each solution of the Schrödinger wave equation is known as an **atomic orbital**. Although the solution of the Schrödinger equation for any system containing more than one electron requires the iterative techniques available to computers, it may be solved for the hydrogen atom (and for **hydrogen-like atoms** such as He^+, Li^{2+}, ...) by analytical means, the molar energy solutions being represented by the equation:

$$E_n = -\frac{N_A \mu Z^2 e^4}{8\varepsilon_0^2 h^2}\left[\frac{1}{n^2}\right] \quad (2.3)$$

where Z is the atomic number (equal to the number of protons in the nucleus), μ is the reduced mass of the system, defined by the equation:

$$\frac{1}{\mu} = \frac{1}{m_e} + \frac{1}{m_n} \qquad (2.4)$$

in which m_e is the mass of the electron and m_n is the mass of the nucleus, e is the electronic charge, and ε_0 is the **permittivity** (dielectric constant in older terminology) of a vacuum ($8.8541878 \times 10^{-12}$ F m^{-1}). This mixture of units is equivalent to J mol^{-1}. [To check that statement it is essential to know the relations: 1 C = 1 A s; 1 F (farad) = 1 A^2 s^4 kg^{-1} m^{-2} (A = ampere); 1 J = 1 kg m^2 s^{-2}.]

The reduced mass takes into account that the system of nucleus plus electron has a centre of mass which does not coincide with the centre of the nucleus. In classical physics the two particles revolve around the centre of mass.

Worked Problem 2.1

Q Calculate the value of the reduced mass of the hydrogen atom.

A Equation (2.4) may be rearranged to read:

$$\mu = \frac{m_e m_p}{m_e + m_p}$$

where m_p is the mass of the proton. Using the values from Table 1.1, the value of μ is: $9.1044313 \times 10^{-31}$ kg to eight significant figures.

By comparing equations (2.3) and (2.1), with a Z value of 1, it may be concluded that the value of the Rydberg constant is given by:

$$R = \frac{\mu e^4}{8\varepsilon_0^2 ch^3} \qquad (2.5)$$

Equation (2.3) gives the permitted energies for the electron in the hydrogen atom. The value of the Rydberg constant given by equation (2.5) is identical to the observed value obtained from spectroscopic measurements of the hydrogen atom emission lines.

Worked Problem 2.2

Q Using the values of the universal constants and the value of the reduced mass of the hydrogen atom previously calculated, calculate the value of the Rydberg constant and compare it with the experimental value quoted in Section 1.6.

A The answer is 1.0967758×10^7 m^{-1} to eight significant figures. The units of the universal constants, given above, reduce to reciprocal metres. An excellent agreement with the experimental value.

The solutions of the Schrödinger equation show how ψ is distributed in the space around the nucleus of the hydrogen atom. The solutions for ψ are characterized by the values of three quantum numbers (essentially there are three because of the three spatial dimensions, x, y and z), and every allowed set of values for the quantum numbers, together with the associated wave function, describes what is termed an **atomic orbital**. Other representations are used for atomic orbitals, such as the boundary surface and other diagrams described later in the chapter, but the strict definition of an atomic orbital is its mathematical wave function.

2.1.3 The Quantum Rules and Atomic Orbitals

The **quantum rules** are statements of the permitted values of the quantum numbers, n, l and m.

(i) The **principal quantum number**, n (the same n as in equation 1.16), has values that are integral and non-zero:

$$n = 1, 2, 3, 4, \ldots$$

It defines groups of orbitals which are distinguished, within each group, by the values of l and m_l.

(ii) The **secondary or orbital angular momentum quantum number**, l, as its name implies, describes the orbital angular momentum of the electron, and has values which are integral, including the value zero:

$$l = 0, 1, 2, 3, \ldots (n - 1)$$

For a given value of n, the maximum permitted value of l is $(n - l)$, so that for a value of $n = 3$, l is restricted to the values 0, 1 or 2.

Worked Problem 2.3

Q If the value of n is 5, what are the permitted values of l?

A The permitted values of l are 0, 1, 2, 3 and 4.

(iii) The **magnetic quantum number**, m_l, so called because it is related to the behaviour of electronic energy levels when subjected to an external magnetic field, has values which are dependent upon the value of l. The permitted values are:

$$m_l = +l, +l - 1, \ldots 0, \ldots -(l - 1), -l$$

For instance, a value of $l = 2$ would yield five different values of m_l: 2, 1, 0, –1 and –2. In general there are $2l + 1$ values of m_l for any given value of l. In the absence of an externally applied magnetic field, the orbitals possessing a given l value would have identical energies. Orbitals of identical energy are described as being **degenerate**. In the presence of a magnetic field the **degeneracy** of the orbitals breaks down; they have different energies. The breakdown of orbital degeneracy (for a given l value) in a magnetic field explains the **Zeeman effect**. This is the observation that in the presence of a magnetic field the atomic spectrum of an element has more lines than in the absence of the field.

Worked Problem 2.4

Q If the value of l is 3, what are the permitted values for m_l?

A The permitted values for m_l are 3, 2, 1, 0, –1, –2 and –3.

Worked Problem 2.5

Q State whether the following sets of quantum number values are valid descriptions of atomic orbitals, and explain why some are invalid.

	n	l	m_l
(a)	2	2	0
(b)	3	1	–1
(c)	3	1	–2

A (a) is invalid because the value of l should be less than that of n; (b) is valid; (c) is invalid because the value of m_l must be within the range $+l$ to $-l$.

The values of l and m_l are dependent upon the value of n, and so it can be concluded that the value of n is concerned with a particular set of atomic orbitals, all characterized by the given value of n. For any one value of n there is the possibility of more than one permitted value of l (except in the case where $n = 1$, when l can only be zero). The notation which is used to distinguish orbitals with different l values consists of a code letter associated with each value. The code letters are shown in Table 2.1.

Table 2.1 Code letters for l values

Value of l	*Code letter*
0	s
1	p
2	d
3	f

Table 2.2 Number of atomic orbitals for a given l value

Value of l	*Number of orbitals*
0	1
1	3
2	5
3	7

Table 2.1 gives only a portion of an infinite set of values of l. Those given are the only values of any interest for the majority of applications to known atoms. The selection of the code letters seems, and is, illogical in that the first four are the initial letters of the words sharp, principal, diffuse and fundamental, words used by 19th century spectroscopists to describe aspects of line spectra. The fifth letter, g, follows on alphabetically, the sixth being h and so on. The numerical value of n and the code letter for the value of l are sufficient for a general description of an atomic orbital. The main differences between atomic orbitals with different l values are concerned with their orientations in space, and those with different n values have different sizes. The main importance of the m_l values is that they indicate the number, $(2l + 1)$, of differently spatially oriented orbitals for the given l value. For instance, if $l = 2$ there are five different values of m_l, corresponding to five differently spatially oriented d orbitals. The number of differently spatially oriented orbitals for particular values of l are given in Table 2.2.

Application of all the above rules allows the compilation of the types of atomic orbital and the number of each type. For $n = 1$, there is a single 1s orbital ($l = 0$, $m_l = 0$). For $n = 2$, there is one 2s orbital ($l = 0$, $m_l = 0$) and three 2p orbitals ($l = 1$, $m_l = 1$, 0 or -1). For $n = 3$, there is one 3s orbital, three 3p orbitals and five 3d orbitals ($l = 2$, $m_l = 2$, 1, 0, -1 or -2). For $n = 4$, there is one 4s orbital, three 4p orbitals, five 4d orbitals and seven 4f orbitals ($l = 3$, $m_l = 3$, 2, 1, 0, -1, -2 or -3), and so on up to and beyond the normally practical limit of $n = 7$.

Worked Problem 2.6

Q Which atomic orbitals are represented by the following combinations of quantum number values:

(a) $n = 4$, $l = 3$ and $m_l = 2$
(b) $n = 2$, $l = 0$ and $m_l = 0$
(c) $n = 5$, $l = 2$ and $m_l = -2$

A (a) A 4f orbital; (b) the 2s orbital; and (c) a 5d orbital.

The sequence can be extended to infinity, but in practice it is only necessary to consider values of n up to seven. Not all the orbitals so described are needed for the electrons in known atoms in their ground electronic states. Higher values of the n quantum number are required to interpret the emission spectra of the heavier elements.

For every increase of one in the value of n there is an extra type of orbital. The number of atomic orbitals associated with any value of n is given by n^2, so that for $n = 1$ there is one orbital (1s), for $n = 2$ there are four orbitals (one 2s and three 2p), for $n = 3$ there are nine orbitals (one 3s, three 3p and five 3d), and for $n = 4$ there are 16 orbitals (one 4s, three 4p, five 4d and seven 4f).

For the hydrogen atom, the orbitals having a particular value of n all have the same energy: they are *degenerate*. Level one (n = 1) is singly degenerate (*i.e.* non-degenerate), level two (n = 2) has a degeneracy of four, and so on. The degeneracy of any level (*i.e.* the number of orbitals with identical energy) is given by the value of n^2. Such widespread degeneracy of electronic levels in the hydrogen atom is the basis of the simplicity of the diagram of the levels shown in Figure 1.7. The orbital energies are determined solely by the value of n in equation (1.23).

The emission spectrum of helium, an atom containing two electrons, is considerably more complex than that of the hydrogen atom. There are about twice as many lines in the visible region than are found in that region in the hydrogen spectrum.

2.1.4 Spatial Orientations of Atomic Orbitals

The spatial orientations of the atomic orbitals of the hydrogen atom are very important in the consideration of the interaction of orbitals of different atoms in the production of chemical bonds. The solutions of the Schrödinger wave equation for the hydrogen atom may be represented by the equation:

$$\psi = RA \tag{2.6}$$

where ψ is the wave function for an atomic orbital of the hydrogen atom, R represents the radial function (the manner in which ψ varies along *any* line radiating from the nucleus) and A is the angular function which takes into account variations of ψ dependent upon the particular direction of a line radiating from the nucleus with respect to the coordinate axes.

Box 2.1 Coordinate Axes and Polar Coordinates

Atomic orbital wave functions are commonly expressed in terms of polar coordinates, rather than Cartesian ones. Figure 2.1 shows the relationship between Cartesian (x, y, z) and polar (r, θ, ϕ) coordinates. The values of x, y and z in terms of polar coordinates are given by:

$$x = r\sin\theta\cos\phi$$
$$y = r\sin\theta\sin\phi$$
$$z = r\cos\theta$$

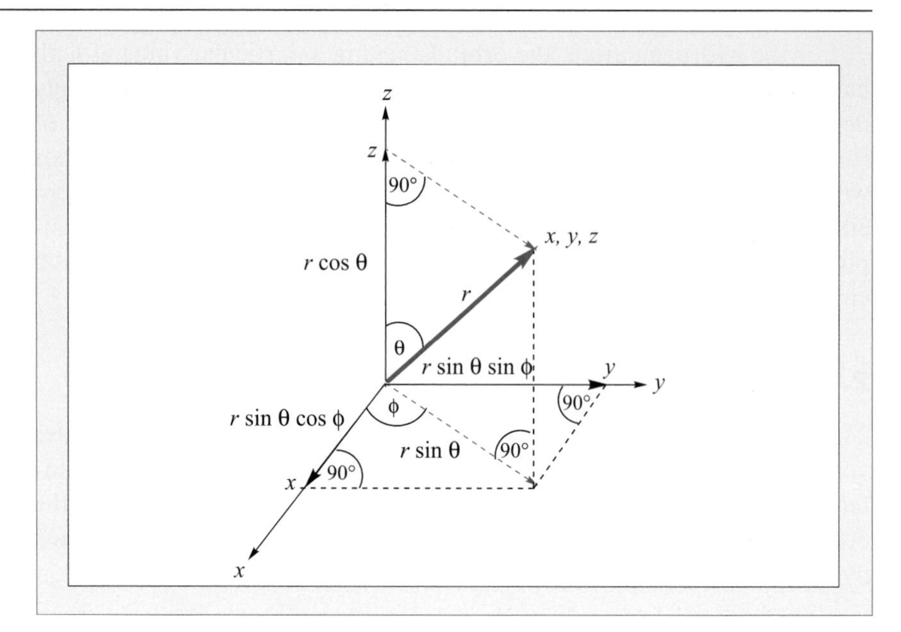

Figure 2.1 A diagram showing the relationship between Cartesian coordinates (x, y, z) and polar coordinates (r, θ, ϕ)

As examples of radial and angular wave functions, those for values of the principal quantum number, n, up to 3 are given, respectively, in Tables 2.3 and 2.4. Z represents the atomic number (1 for the hydrogen atom, but the formulae shown represent hydrogen-like atoms such as He^+ for which $Z = 2$), and the term a_0 is the atomic unit of distance, explained below, and known as the Bohr radius. It has the value 52.9177 pm.

Table 2.3 The radial functions of the 1s, 2s, 2p, 3s, 3p and 3d atomic orbitals; $\rho = 2Zr/a_0$

Atomic orbital	*R*
1s	$2\left(\frac{Z}{a_0}\right)^{3/2} e^{-\rho/2}$
2s	$\frac{1}{2(2)^{1/2}}\left(\frac{Z}{a_0}\right)^{3/2}(2-\frac{1}{2}\rho)e^{-\rho/4}$
2p	$\frac{1}{4(6)^{1/2}}\left(\frac{Z}{a_0}\right)^{3/2}\rho e^{-\rho/4}$
3s	$\frac{1}{9(3)^{1/2}}\left(\frac{Z}{a_0}\right)^{3/2}(6-2\rho+\frac{1}{9}\rho^2)e^{-\rho/6}$
3p	$\frac{1}{27(6)^{1/2}}\left(\frac{Z}{a_0}\right)^{3/2}(4-\frac{1}{3}\rho)\rho e^{-\rho/6}$
3d	$\frac{1}{81(30)^{1/2}}\left(\frac{Z}{a_0}\right)^{3/2}\rho^2 e^{-\rho/6}$

Table 2.4 The angular functions for the 1s, 2s, 2p, 3s, 3p and 3d atomic orbitals

Value of l	*Value of m_l*	A^a
0	0	$(1/4\pi)^{1/2}$
1	0	$(3/4\pi)^{1/2}\cos\theta$
1	±1	$(3/8\pi)^{1/2}\sin\theta\, e^{\pm i\phi}$
2	0	$(5/16\pi)^{1/2}(3\cos^2\theta - 1)$
2	±1	$(15/8\pi)^{1/2}\cos\theta \sin\theta\, e^{\pm i\phi}$
2	±2	$(15/32\pi)^{1/2}\sin^2\theta\, e^{\pm 2i\phi}$

[a] The exponential terms containing *i* may be converted into trigonometric terms by using Euler's identities (themselves derived from the theory of infinite series): $e^{i\phi} = \cos\phi + i\sin\phi$; $e^{-i\phi} = \cos\phi - i\sin\phi$; $e^{2i\phi} = \cos2\phi + i\sin2\phi$; $e^{-2i\phi} = \cos2\phi - i\sin2\phi$

In Table 2.4 the form of the angular function, *A*, is dependent upon the value of the magnetic quantum number, m_l, in addition to the value of *l*; when the value of *l* is greater than zero the angular functions contain the term *i*, the square root of minus 1.

Box 2.2 Atomic Orbitals and m_l Values

Table 2.4 is included in the text to make the important point that the atomic orbitals which are generally used by chemists are ones which are *real* functions, *i.e.* they do not contain *i*. In order to obtain such functions it is necessary to take linear combinations of the angular functions containing *i*, so that the resulting functions are real and can then be visualized.

The 2p orbitals for which the values of m_l are 1 and –1 respectively are combined in two linear combinations so that the orbitals do not have imaginary parts and can be visualized diagrammatically. The $2p(m_l = 1) + 2p(m_l = -1)$ combination leads to the $2p_x$ orbital and the $2p(m_l = 1) - 2p(m_l = -1)$ combination leads to the $2p_y$ orbital. By similar procedures, the 3p orbitals for which the values of m_l are 1 and –1, respectively, become the $3p_x$ and $3p_y$ orbitals, and the 3d orbitals with non-zero values of m_l become the $3d_{xz}$ $(d_1 + d_{-1})$, $3d_{yz}$ $(d_1 - d_{-1})$, $3d_{x^2-y^2}$ $(d_2 + d_{-2})$ and $3d_{xy}$ $(d_2 - d_{-2})$ orbitals.

The 1s atomic orbital of the hydrogen atom is spherically symmetrical and so the angular function is a constant term [$A_{1s} = 1/(4\pi)^{1/2}$; see Table 2.4] chosen to normalize the wave function so that the integral of its square over all space has the value +1, as expressed by the general equation:

$$\int_0^\infty \psi^2 \mathrm{d}\tau = 1 \tag{2.7}$$

Normalization consists of equating the mathematical *certainty* of finding the electron somewhere in the orbital to a value +1. The radial function is given by the equation:

$$R_{1s} = 2a_0^{-3/2}e^{-r/a_0} \tag{2.8}$$

where r is the distance from the nucleus, and a_0 is the atomic unit of length consisting of a collection of universal constants and given by:

$$a_0 = h^2\varepsilon_0/\pi me^2 \tag{2.9}$$

where h is Planck's constant, ε_0 is the permittivity of a vacuum, m is the mass of the electron and e is the electronic charge (the units of equation 2.9 combine to yield m^{-1}).

The quantity, a_0, is referred to as the **Bohr radius** (52.917726 pm) because it is identical to the radius of the orbit of the 1s electron in the "Bohr atom". The early Bohr theory of the atom invoked electron **orbits** of *definite* radii, but these are invalid since they violate the Heisenberg uncertainty principle. The Bohr radius is used as the **atomic unit of length**.

Figure 2.2 shows a plot of the 1s wave function against distance from the hydrogen atom nucleus. It is a maximum at the nucleus and falls away exponentially with distance, and is asymptotic to the distance axis, getting ever closer to zero value of ψ, but becoming zero only at infinite distance.

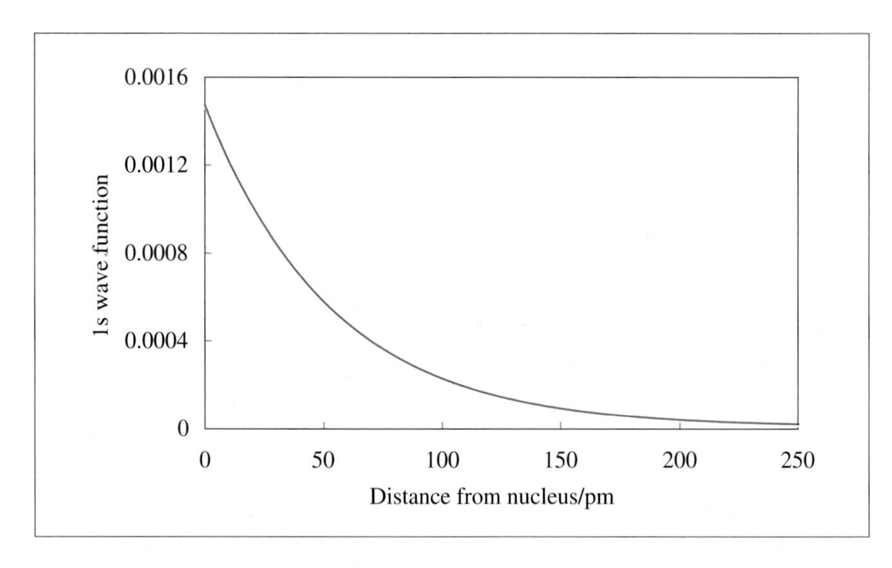

Figure 2.2 A plot of the hydrogen 1s wave function against the distance from the nucleus

To make use of the wave function in a meaningful manner it is necessary to express it in terms of the **radial distribution function**, RDF. This is the variation with distance from the nucleus of the function $4\pi r^2\psi_{1s}^2$, which takes into account the probability of finding an electron between two spheres of respective radii r and $r + \mathrm{d}r$, where $\mathrm{d}r$ is an infinitesimally small radius increment. The equation for this is:

$$\mathrm{RDF}_{1s} = 4\pi r^2 \psi_{1s}^{\ 2} = 4r^2 a_0^{\ -3} e^{-2r/a_0} \qquad (2.10)$$

A plot of RDF_{1s} against r is shown in Figure 2.4, and indicates that the maximum probability of finding the electron is at a distance a_0 (= 52.917726 pm) from the nucleus. One common pictorial representation of atomic orbitals is the solid figure, or **boundary surface**, in which there is a 95% chance (a probability of 0.95) of finding the electron. The 95% boundary surface for the 1s atomic orbital of hydrogen is shown in Figure 2.5. It has a radius of 160 pm.

Figure 2.3 represents two spheres with radii r and $r + dr$. The difference in volumes of the two spheres is given by:
$\frac{4}{3}\pi(r + dr)^3 - \frac{4}{3}\pi r^3$
$= \frac{4}{3}\pi r^3 + 4\pi r^2 dr + 4\pi r(dr)^2 + \frac{4}{3}\pi(dr)^3 - \frac{4}{3}\pi r^3$
$= 4\pi r^2 dr$ (ignoring the squared and cubed terms of dr)

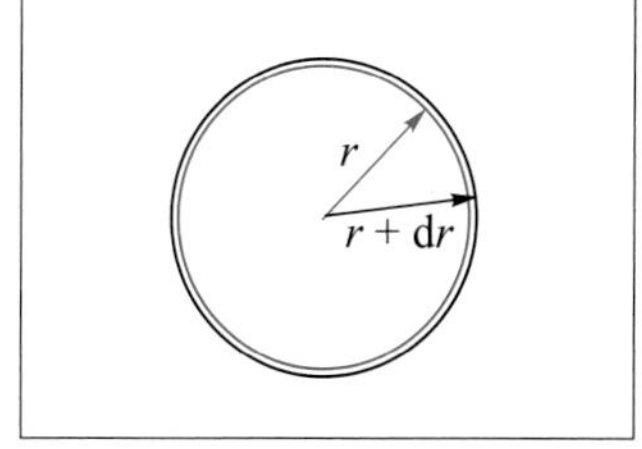

Figure 2.3 A diagram showing how to estimate the volume element $4\pi r^2 dr$ between two spheres of radii r and $r + dr$

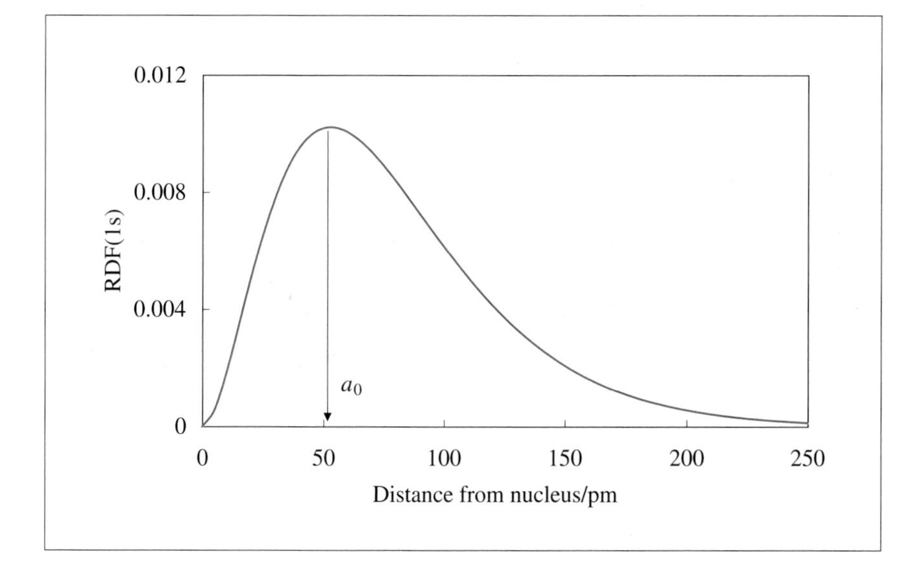

Figure 2.4 The radial distribution function for the 1s atomic orbital of the H atom

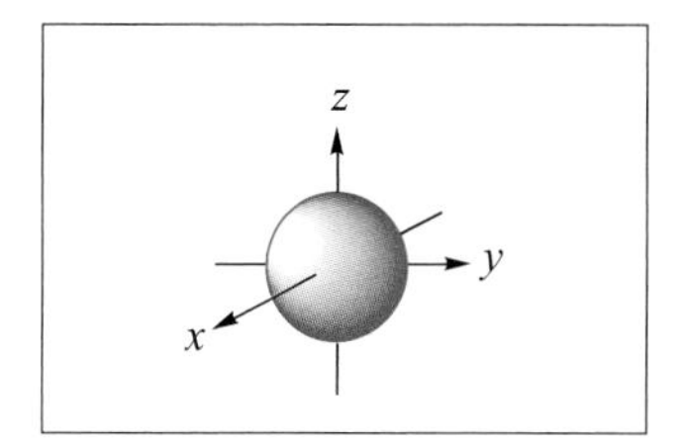

Figure 2.5 The 95% boundary surface for the 1s atomic orbital of hydrogen

Exercise 2.1

Use a spreadsheet or a program such as *Mathematica* to explore the mathematical forms of some of the atomic orbitals described in Tables 2.3 and 2.4. It is not necessary to program the whole of an equation into the spreadsheet, just the portion which varies with r, θ and ϕ; the normalizing factors may be ignored as they are merely multipliers of the functions. It is suggested that you compare the radial functions of the 1s, 2s and 3s orbitals. The 1s diagram is given in Figure 2.4, and you should find that the 2s plot is of high value at the nucleus, but then falls off exponentially and even becomes negative and then rises asymptotically (*i.e.* approaches at infinite distance) to the r axis. The change between positive and negative ψ values indicates the presence of a **radial node**. The plot for the 3s orbital exhibits two radial nodes.

The radial plots of the 1s, 2s and 3s orbital wave functions can easily be transformed into radial distribution plots by multiplying the value of ψ^2 by $4\pi r^2$ and plotting the results against r. The positions of any radial nodes are shown by where the functions cross the x axis, and it should be noticed how the maximum value for the radial distribution function increases with the value of the principal quantum number.

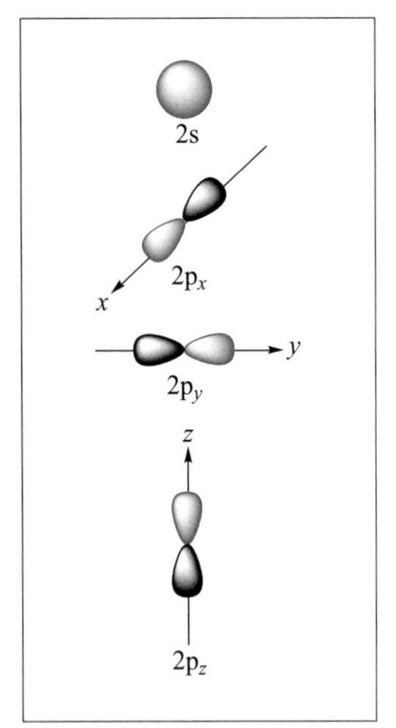

Figure 2.6 The envelope diagrams of the 2s and 2p atomic orbitals of the H atom

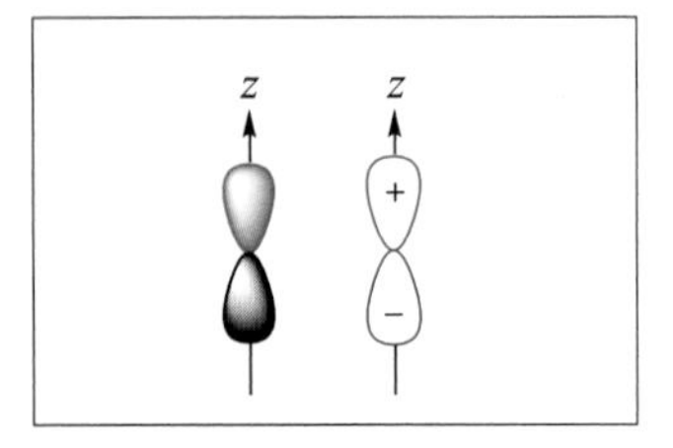

Figure 2.7 Two methods of representing the 2p orbitals

Although the formal method of describing orbitals is to use mathematical expressions, much understanding of orbital properties may be gained by the use of pictorial representations. The most useful pictorial representations of atomic orbitals are similar to boundary surfaces (which are based on ψ^2) but are based upon the distribution of ψ values, with the sign of ψ being indicated in the various parts of the diagram. The shapes of these distributions are based upon the contours of ψ within which the values of ψ^2 represent 0.95, and may be called **orbital envelope diagrams**. The orbital envelopes for the 2s and 2p hydrogen orbitals are shown in Figure 2.6.

Two methods are used to represent the positive and negative values of the wave functions in **atomic orbital envelope diagrams**. Both are shown in Figure 2.7 for the $2p_z$ orbital.

The two lobes of the orbital may be indicated by open areas containing the signs of the wave function in the two areas, or they may be represented by filled and open areas which represent the positive and negative values of the wave function, respectively. In the remainder of the book the filled and open parts of atomic orbital diagrams are used to denote positive and negative signs of wave functions.

Showing the 2s orbital with an everywhere-positive atomic orbital diagram is a matter of convention. The result of the exercise of plotting the wave function for the orbital shows that there is a radial node with the higher values of r yielding negative values of ψ. The radial distribution function indicates that the major probability of finding the electron is in the region of negative ψ values. Nevertheless, it is conventionally depicted as being everywhere positive, this being important only when considering the interaction of the orbital with an orbital of another atom. In such cases, the respective signs of ψ of the orbitals is crucial to whether the atoms bind to each other or otherwise. That there is a **radial node** (a node is where there is a change of sign of ψ) in the radial wave function of the 2s orbital is indicated (see Table 2.3) in the wave function by the term $(2 - \frac{1}{2}\rho)e^{-\rho/4}$. As ρ increases, the term in brackets becomes more and more important, and R becomes negative when $\frac{1}{2}\rho$ is greater than 2. After a minimum is reached, ψ becomes less negative only because the exponential multiplier is becoming excessively small.

A 2p orbital has a nodal surface (*e.g.* the xy plane in the case of the $2p_z$ orbital), and in general orbitals have l such surfaces. Atomic orbital wave functions have $n - l - 1$ radial nodes (nodal spheres) plus l nodal surfaces which pass through the origin, making a total of $n - 1$ nodes.

The three 2p orbitals are directed along the x, y and z axes and are described respectively as $2p_x$, $2p_y$ and $2p_z$. They each consist of two lobes, one of which has positive ψ values, the other having negative ψ values. The envelopes of the five 3d orbitals are shown in Figure 2.8. Those of the seven 4f orbitals are shown in Figure 2.9.

The subscript descriptions of the d orbitals are chosen because such expressions are an important part of their respective mathematical formulations, representing the Cartesian component of the angular wave function, just as the 2p orbitals have the subscripts x, y and z.

It is generally accepted that the orbitals of polyelectronic atoms have spatial distributions similar to those of the hydrogen atom. The spatial orientations and the *signs* of ψ are extremely important in the understanding of chemical bonding. For example, the $3d_{x^2-y^2}$ orbital is a linear combination of the d orbitals which have m_l values of 2 and –2, and contains the angular term $r^2\sin^2\phi\cos2\phi$. This may be manipulated in the following manner:

$$\begin{aligned} r^2\sin^2\phi\cos2\phi &= r^2\sin^2\phi(\cos^2\phi - \sin^2\phi) \\ &= r^2\sin^2\phi\cos^2\phi - r^2\sin^2\phi\sin^2\phi \\ &= x^2 - y^2 \end{aligned}$$

the Cartesian version being used as the designatory subscript for the orbital.

The f orbital designatory subscripts, similarly to those of the d orbitals, are x^3, y^3, z^3, $x(y^2 - z^2)$, $y(z^2 - x^2)$, $z(x^2 - y^2)$ and xyz, as indicated in Figure 2.9.

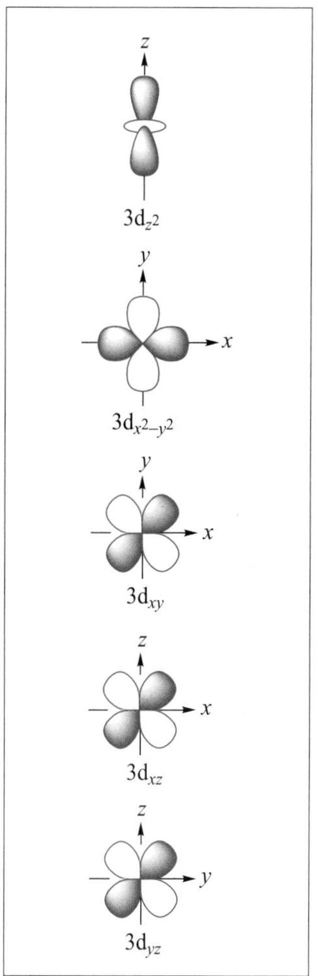

Figure 2.8 The envelopes of the five 3d orbitals

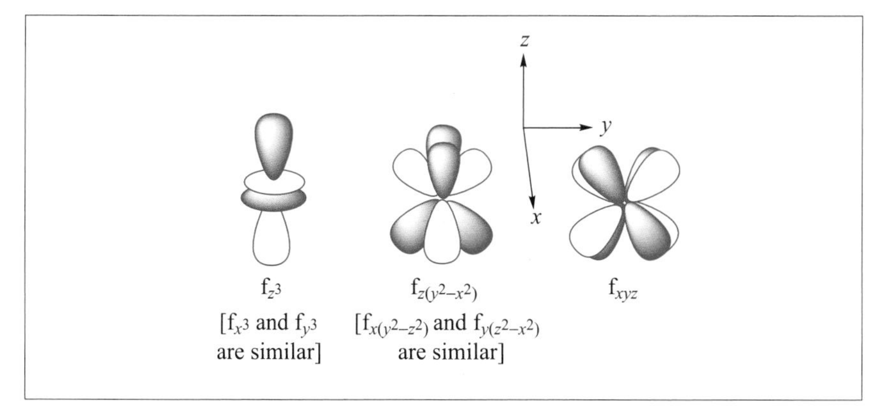

Figure 2.9 The envelopes of the seven 4f orbitals

Summary of Key Points

1. The energy levels of the hydrogen atom were described, and shown to be dependent solely upon the value of the principal quantum number, n.

2. The necessity for quantum mechanical calculations was emphasized because of the large uncertainty in knowledge of the position of an electron in the hydrogen atom, the uncertainty in energy being relatively small.

3. Brief introductions to quantum mechanics and the Schrödinger equation were given. The solutions of the equation were shown to contain information regarding the energies of the permitted levels and the mathematical forms of the associated wave functions. The wave functions accurately describe atomic orbitals.

4. The quantum rules were described in terms of the permitted values of the three quantum numbers n, l and m_l.

5. The spatial orientations of the atomic orbitals of the hydrogen atom were described in terms of atomic orbital diagrams, based upon envelopes of ψ values.

Problems

2.1. In a hydrogen atom, what are the degeneracies of the 3p, 4d and 5d atomic orbitals?

2.2. For any value of n greater than 2, show that there can only be five d orbitals.

2.3. For any value of n greater than 3, show that there can only be seven f orbitals.

2.4. Determine the values of r at which there are radial nodes in the 3s atomic orbital. Solve the equation $6 - 2\rho + {}^{1}/_{9}\rho^2 = 0$ to achieve the result, given that $\rho = 2Zr/a_0$. Compare them with your graph of the radial distribution function for this orbital.

Further Reading

D. O. Hayward, *Quantum Mechanics for Chemists*, Royal Society of Chemistry, Cambridge, 2002. A companion volume in this series.

3

The Electronic Configurations of Atoms; the Periodic Classification of the Elements

The chemistry of an element is determined by the manner in which its electrons are arranged in the atom. Such arrangements and their chemical consequences are the subject of this chapter, leading to a general description of the structure of the modern periodic classification of the elements: the Periodic Table.

Aims

By the end of this chapter you should understand:

- That interelectronic repulsion is responsible for the complex nature of polyelectronic atoms
- That hydrogen-like atomic orbitals suffer a loss of degeneracy when more than one electron is present
- That electrons possess an amount of intrinsic energy governed by the spin quantum number
- The Pauli exclusion principle
- That the numbers of atomic orbitals in an atom are dependent upon the Pauli exclusion principle
- That a maximum of two electrons may occupy an atomic orbital
- That electrons are indistinguishable from one another
- The *aufbau* principle and the order of filling the available atomic orbitals
- Hund's rules
- The general structure of the Periodic Table
- The irregularities in the filling of sets of d and f orbitals

3.1 Polyelectronic Atoms

The treatment of atoms with more than one electron – polyelectronic atoms – requires consideration of the effects of interelectronic repulsion, orbital penetration towards the nucleus, nuclear shielding, and an extra quantum number – the spin quantum number – which specifies the intrinsic energy of the electron in any orbital. The restriction on the numbers of atomic orbitals and the number of electrons that they can contain leads to a discussion of the Pauli exclusion principle, Hund's rules and the *aufbau* principle. All these considerations are necessary to allow the construction of the modern form of the periodic classification of the elements.

3.1.1 Interelectronic Repulsion

Deviations from the relative simplicity of one-electron atoms arise in atoms which contain more than one electron: polyelectronic atoms. Even an atom possessing two electrons is not treatable by the analytical mathematics which produced the solution of the wave equation for the hydrogen atom, as described in Chapter 2. Coulombic interelectronic repulsion is of great importance and is a subject not given sufficient emphasis in most pre-university courses. Negatively charged electrons repel each other, although they can pair up under certain conditions, and the understanding of the effects of such repulsion is fundamental to the rationalization of a great amount of chemistry. The wave equation (2.2) can only be solved for polyelectronic systems by using the iterative capacity of computers. The basis of the method is to guess the form of ψ for each orbital employed and to calculate the corresponding energy of the atomic system. The atomic orbitals that are acceptable as solutions of the wave equation are those which confer the minimum energy upon the system. The solutions, in general, mimic those for the hydrogen atom, and the same nomenclature may be used to describe the orbitals of polyelectronic systems as is used for the hydrogen atom. The shapes of their spatial distributions are similar to those for hydrogen. The major difference arises in the energies of the orbitals. The degeneracy of the levels with a given n value is lost. Orbitals with the same n value and with the same l value are still degenerate. Those with the same n value but with different values of l are no longer degenerate. This does not affect the 1s orbital, which is singly degenerate, but it does affect the 2s and 2p orbitals. The 2p orbitals have a higher energy than do the 2s, the three 2p orbitals retaining their three-fold degeneracy. Likewise, the sets of orbitals with higher values of n split into s, p, d, f, ... sub-sets which within themselves retain their degeneracy (given by the value of $2m_l$ + 1). In general, the energy of an atomic orbital decreases with the square of the nuclear charge, as indicated by equation (2.1). In polyelectronic atoms the effects of interelectronic repulsion are superimposed upon this trend. The loss of degeneracy, together with the general decrease in energy as Z increases, causes changes in the order of energies of various sub-sets of atomic orbitals, as is discussed in detail in Section 3.3.

3.1.2 Orbital Penetration Effects

The reasons for the loss of degeneracy of sets of atomic orbitals with the same value of n are embedded in the radial distribution functions of the orbitals. The general effect in polyelectronic atoms is for the degeneracy of a set of atomic orbitals with a given n value to break up into sub-sets such that the s orbital is of lower energy than the p orbitals, the p orbitals are lower than the d orbitals, and the d orbitals are lower than the f orbitals, as shown in Figure 3.1.

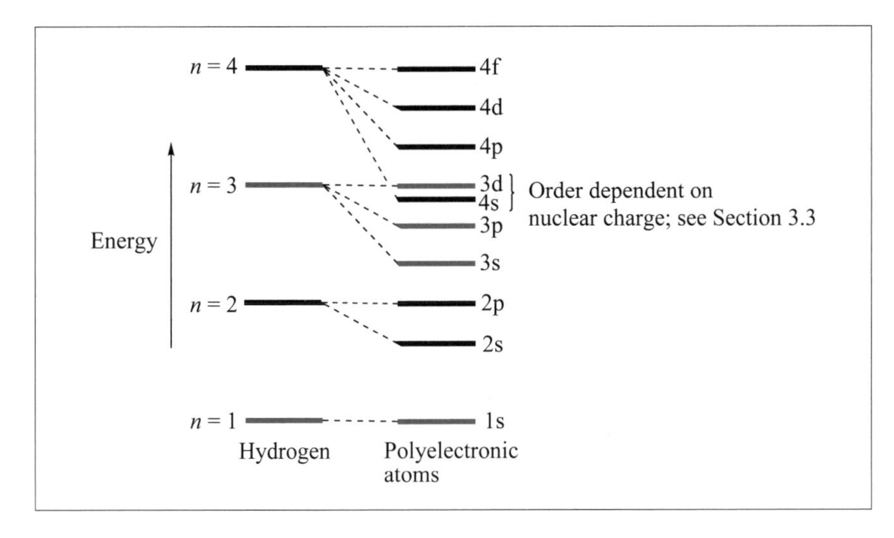

Figure 3.1 A diagram showing the breakdown of degeneracy in polyelectronic atoms (not to scale)

Figure 3.2 shows the radial distribution functions for the hydrogen 2s and 2p orbitals, from which it can be seen that the 2s orbital has a considerably larger probability near the nucleus than the 2p orbital. When an electron in a polyelectronic atom occupies the $n = 2$ level, it would be more stable in the 2s orbital than in the 2p orbital. In the 2s orbital it would be nearer the nucleus and be more strongly attracted than if it were to occupy the 2p sub-set.

Similar considerations explain the breakdown of the degeneracies of sets of orbitals with a given n value into their sub-sets, *i.e.* s, p, d, *etc.*

3.1.3 Nuclear Shielding

Another aspect of the interaction of electrons in polyelectronic atoms is

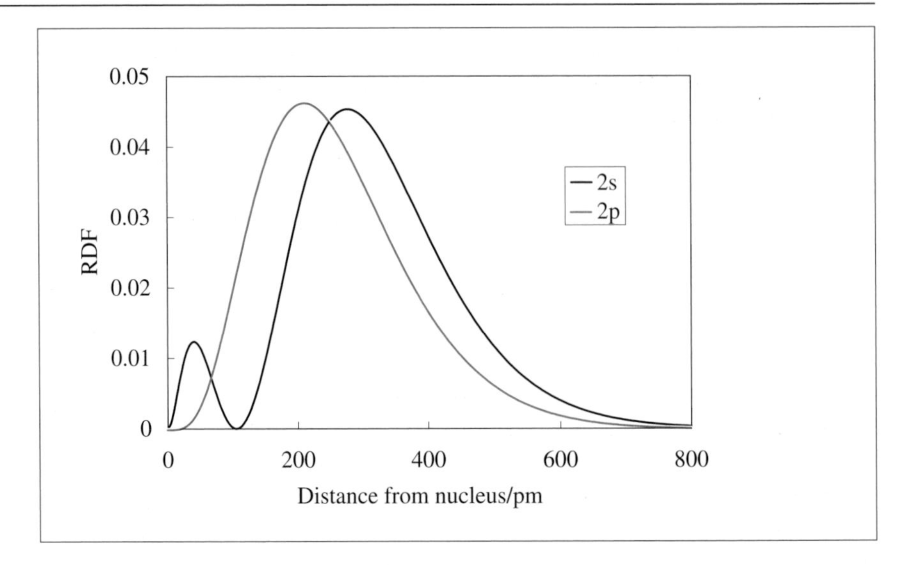

Figure 3.2 The radial distribution functions (RDF) for the hydrogen 2s and 2p orbitals

that of nuclear shielding. Any electron is held in its atomic orbital by the Coulombic attraction of the nuclear charge, but when another electron is present the two electrons repel each other to destabilize the system. This repulsion may be interpreted in terms of one electron shielding the other electron from the effect of the nucleus. This leads to the concept of the **effective nuclear charge**, Z_{eff}, which is that experienced by one electron as the result of the presence in the atom of other electrons. By its better penetration towards the nucleus, an electron in a 2s orbital is less shielded than an electron in a 2p orbital. The combination of penetration and shielding ensures the breakdown of degeneracy of orbitals of equal n values.

An important aspect of shielding applies to the 2p, 3d and 4f orbitals, in which electrons are strongly affected by the nuclear charge because there are no 1p, 2d and 3f orbitals (see Section 2.1.3) closer to the nucleus that can accommodate shielding electrons with the same spatial distributions. In turn, electrons in the 2p, 3d and 4f orbitals are very efficient at shielding electrons in the 3p, 4d and 5f orbitals. This factor is mainly responsible for the special characteristics of the elements of the second period; the elements show considerable differences in properties from those of the other members of their groups. It is also a major factor in ensuring that the properties of the elements of the first transition series differ from those of the corresponding elements in the second and third series.

3.1.4 The Spin Quantum Number

When dealing with atoms possessing more than one electron it is necessary to consider the electron-holding capacity of the orbitals of that atom. In order to do this it becomes necessary to introduce a fourth quantum number: s, the **spin quantum number**. This is concerned with the quantized amount of energy possessed by the electron, independent of that concerned with its passage around the nucleus, the latter energy being controlled by the value of l. The electron has an intrinsic energy which is associated with the term spin. This is unfortunate, since it may give rise to the impression that electrons are spinning on their own axes much as the Moon spins on its axis with a motion which is independent of its orbital motion around the Earth. The uncertainty principle indicates that observation of the position of an electron is impossible, so that the visualization of an electron spinning around its own axis must be left to the imagination! However, to take into account the intrinsic energy of an electron, the value of s is taken to be ½. Essentially, the intrinsic energy of the electron may interact in a quantized manner with that associated with the angular momentum represented by l, such that the only permitted interactions are $l + s$ and $l - s$. For atoms possessing more than one electron it is necessary to specify the values of s with respect to an applied magnetic field; these are expressed as values of m_s of +½ or –½ , *i.e.* $m_s = s$ or $s - 1$.

The best known example of the effect of the interaction of orbital and spin momenta is the origin of the characteristic yellow emission from sodium atoms when they are electronically excited, as they are in street lights, and in gas flames when the cooking boils over! The sodium atom has a 3s electron (l = 0), which may be excited to the 3p level in which l = 1 and with which s can interact in a quantized way, $l + s$ and $l - s$, to give two possible energies for the excited state. When these differently excited atoms return to the ground state, two different (yellow) photons are emitted, with wavelengths of 589.59 and 588.95 nm. The changes are shown in Figure 3.3. The spin momentum of the ground state is unable to interact with the zero orbital momentum, so there is no splitting of the ground state as there is of the excited state.

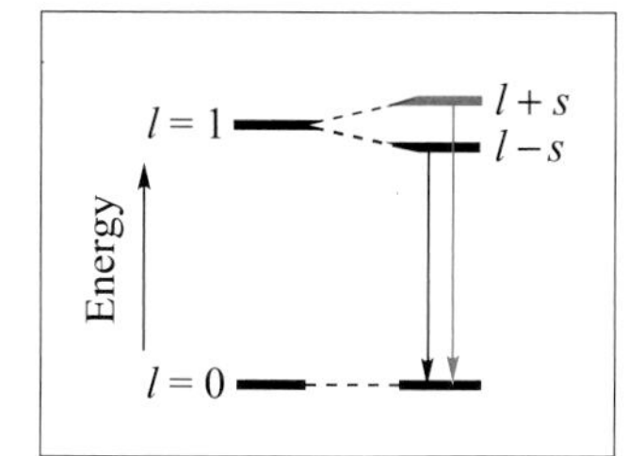

Figure 3.3 An orbital explanation of the sodium doublet emission giving evidence for electron spin

3.1.5 The Pauli Exclusion Principle

The simple conclusion that *the maximum number of electrons which may occupy any orbital is two* arises from the **Pauli exclusion principle**. This is the cornerstone in understanding the chemistry of the elements. It may be stated as:

"No two electrons in an atom may possess identical sets of values of the four quantum numbers, n, l, m_l and m_s."

The consequences are:

(i) to restrict the number of electrons per orbital to a *maximum of two*, and

(ii) to restrict any one atom to only *one* particular orbital, defined by its set of n, l and m_l values.

Consider the 1s orbital; $n = 1$, $l = 0$ and $m_l = 0$, and there are no possibilities for changes in these values (any electron in a 1s orbital must be associated with them). One electron could have a value of m_s of ½ , but the second electron must have the alternative value of m_s of –½ . The two electrons occupying the same orbital must have opposite "spins". Since there are no other combinations of the values of the four quantum numbers, it is concluded that only two electrons may occupy the 1s orbital and that there can only be one 1s orbital in any one atom. Similar conclusions are valid for all other orbitals. The application of the Pauli exclusion principle provides the necessary framework for the observed electronic configurations of the elements.

Worked Problem 3.1

Q Why can there be only a maximum of six 2p electrons in one atom?

A The 2p orbitals have $n = 2$ and $l = 1$, there can only be variations in the values of m_l and m_s. There are only six different combinations for these quantum numbers as shown below:

m_l	m_s
1	$\pm\frac{1}{2}$
0	$\pm\frac{1}{2}$
−1	$\pm\frac{1}{2}$

Thus there can only be a maximum of six 2p electrons in one atom without violating the Pauli principle.

The orbital wave functions which have been discussed above do not represent the total wave functions of the electrons. The total wave function must include a spin wave function, ψ_{spin}, and can be written as:

$$\psi_{total} = \psi_{orbital}\psi_{spin} \tag{3.1}$$

For two electrons, labelled 1 and 2 and residing singly in two orbitals labelled *a* and *b*, taking into account the **indistinguishability of electrons**, there are two possible total wave functions which are written as:

$$\psi_m = [\psi_a(1)\psi_b(2) + \psi_a(2)\psi_b(1)]\psi_{spin} \tag{3.2}$$

$$\psi_n = [\psi_a(1)\psi_b(2) - \psi_a(2)\psi_b(1)]\psi_{spin} \tag{3.3}$$

That electrons are *indistinguishable* from each other is very important, and has consequences in the mathematical descriptions of wave functions which must not imply that electrons can be distinguished from each other.

The product $\psi_a(1)\psi_b(2)$ implies that electron 1 is resident in orbital *a* and electron 2 is in orbital *b* and the product $\psi_a(2)\psi_b(1)$ implies that electron 2 is resident in orbital *a* and electron 1 is in orbital *b*. Both products by themselves violate the rule of electron indistinguishability. The linear combinations of the two ways of placing two electrons in the two orbitals represent the method of allowing for the indistinguishability of the two electrons.

Worked Problem 3.2

Q Why is the wave function $\psi_a(1)\ \psi_b(2)$ not valid?

A The wave function implies that electron 1 is resident in orbital *a* and electron 2 is in orbital *b* and that electrons can so be distinguished. This is not possible.

The orbital part of ψ_m is symmetric to electron exchange, *i.e.* it does not change sign if the two electrons are exchanged between their occupancies of the two orbitals. The orbital part of ψ_n *does* change sign if the two electrons are exchanged, and is anti-symmetric to that operation.

If the two electrons possess identical spins, the two spin wave functions can be written as: $\alpha(1)\alpha(2)$ and $\beta(1)\beta(2)$, α and β being used to represent the two possible values of the electron spin. If the two electrons have opposed spins, there are again two spin wave functions which can be written as the linear combinations:

$$\alpha(1)\beta(2) + \alpha(2)\beta(1) \text{ and } \alpha(1)\beta(2) - \alpha(2)\beta(1)$$

to incorporate the indistinguishability of the electrons. The former spin wave function is symmetric with respect to electron exchange and the latter is anti-symmetric to the same operation.

The exchange properties of the orbital and spin wave functions are relevant to an alternative manner of stating the Pauli exclusion principle, which is that the total wave function of a real system must be anti-symmetric. This is a consistent property of all wave functions which are relevant to real systems.

Box 3.1 The Symmetry Properties of Orbital and Spin Wave Functions

The "truth table" or multiplication table for the symmetry properties of orbital and spin wave functions is shown in Table 3.1.

Table 3.1 The truth table for the interactions of the symmetry properties with respect to electron exchange of orbital and spin wave functions

	Spin function	*Symmetric*	*Anti-symmetric*
Orbital function	*Symmetric*	**Symmetric**	**Anti-symmetric**
	Anti-symmetric	**Anti-symmetric**	**Symmetric**

Applied to the two-orbital–two-electron case under consideration, the appropriate total anti-symmetric wave functions are produced by combining symmetric orbital wave functions with anti-symmetric spin wave functions, or by combining anti-symmetric orbital wave functions with symmetric spin wave functions, since the products of two symmetric wave functions or that of two anti-symmetric wave functions are symmetric and would violate the Pauli principle. The appropriate total wave functions are thus:

$$\psi_p = [\psi_a(1)\psi_b(2) + \psi_a(2)\psi_b(1)][\alpha(1)\beta(2) - \alpha(2)\beta(1)] \quad (3.4)$$

$$\psi_q = [\psi_a(1)\psi_b(2) - \psi_a(2)\psi_b(1)][\alpha(1)\beta(2) + \alpha(2)\beta(1)] \quad (3.5)$$

$$\psi_r = [\psi_a(1)\psi_b(2) - \psi_a(2)\psi_b(1)][\alpha(1)\alpha(2)] \quad (3.6)$$

$$\psi_s = [\psi_a(1)\psi_b(2) - \psi_a(2)\psi_b(1)][\beta(1)\beta(2)] \quad (3.7)$$

The orbital state described by equation (3.4), two electrons in separate orbitals with opposed spins, is markedly different from that described by equations (3.5)–(3.7). The latter description, requiring three equations, is of two electrons occupying two separate orbitals and having **parallel spins**. This **triplet state**, represented by the *three* equations, is of lower energy than the singlet state represented by the single equation (3.4). The lower energy of the triplet state arises from the impossibility of the two electrons occupying the same space, as would be the case if they occupied the same orbital. In the latter case, ψ_a is made identical to ψ_b in equations (3.5)–(3.7), and the wave functions have a value of zero. The conclusion is that there is zero probability of the occurrence of the double occupancy of one orbital by two electrons if they possess the same spin. The interelectronic repulsion between two electrons in a triplet state is minimized by comparison with two electrons in a singlet state. In the latter case, if ψ_a is made identical to ψ_b in equation (3.4) the result is non-zero, and indicates that two electrons can occupy the same orbital providing they have opposed spins.

3.2 The Electronic Configurations and Periodic Classification of the Elements

In this section, the principles described in previous sections are applied to the electronic configurations of all the elements, and this leads to a rationalization of the modern form of the **periodic classification of the elements**. The modern form of the **Periodic Table** is shown in Figure 3.4.

The atomic number, Z, of each element is shown together with its electronic configuration. There are 18 groups, according to modern con-

1	2	3	4	5	6	7	8	9	10	11	12	13	14	15	16	17	18
1 **H** $1s^1$																	2 **He** $1s^2$
3 **Li** $2s^1$	4 **Be** $2s^2$											5 **B** $2s^22p^1$	6 **C** $2s^22p^2$	7 **N** $2s^22p^3$	8 **O** $2s^22p^4$	9 **F** $2s^22p^5$	10 **Ne** $2s^22p^6$
11 **Na** $3s^1$	12 **Mg** $3s^2$											13 **Al** $3s^23p^1$	14 **Si** $3s^23p^2$	15 **P** $3s^23p^3$	16 **S** $3s^23p^4$	17 **Cl** $3s^23p^5$	18 **Ar** $3s^23p^6$
19 **K** $4s^1$	20 **Ca** $4s^2$	21 **Sc** $4s^23d^1$	22 **Ti** $4s^23d^2$	23 **V** $4s^23d^3$	24 **Cr** $4s^13d^5$	25 **Mn** $4s^23d^5$	26 **Fe** $4s^23d^6$	27 **Co** $4s^23d^7$	28 **Ni** $4s^23d^8$	29 **Cu** $4s^13d^{10}$	30 **Zn** $4s^23d^{10}$	31 **Ga** $4s^23d^{10}4p^1$	32 **Ge** $4s^23d^{10}4p^2$	33 **As** $4s^23d^{10}4p^3$	34 **Se** $4s^23d^{10}4p^4$	35 **Br** $4s^23d^{10}4p^5$	36 **Kr** $4s^23d^{10}4p^6$
37 **Rb** $5s^1$	38 **Sr** $5s^2$	39 **Y** $5s^24d^1$	40 **Zr** $5s^24d^2$	41 **Nb** $5s^14d^4$	42 **Mo** $5s^14d^5$	43 **Tc** $5s^14d^6$	44 **Ru** $5s^14d^7$	45 **Rh** $5s^14d^8$	46 **Pd** $4d^{10}$	47 **Ag** $5s^14d^{10}$	48 **Cd** $5s^24d^{10}$	49 **In** $5s^24d^{10}5p^1$	50 **Sn** $5s^24d^{10}5p^2$	51 **Sb** $5s^24d^{10}5p^3$	52 **Te** $5s^24d^{10}5p^4$	53 **I** $5s^24d^{10}5p^5$	54 **Xe** $5s^24d^{10}5p^6$
55 **Cs** $6s^1$	56 **Ba** $6s^2$	71 **Lu** $4f^{14}6s^25d^1$	72 **Hf** $4f^{14}6s^25d^2$	73 **Ta** $4f^{14}6s^25d^3$	74 **W** $4f^{14}6s^25d^4$	75 **Re** $4f^{14}6s^25d^5$	76 **Os** $4f^{14}6s^25d^6$	77 **Ir** $4f^{14}6s^25d^7$	78 **Pt** $4f^{14}6s^15d^9$	79 **Au** $4f^{14}6s^15d^{10}$	80 **Hg** $4f^{14}6s^25d^{10}$	81 **Tl** $4f^{14}6s^25d^{10}$ $6p^1$	82 **Pb** $4f^{14}6s^25d^{10}$ $6p^2$	83 **Bi** $4f^{14}6s^25d^{10}$ $6p^3$	84 **Po** $4f^{14}6s^25d^{10}$ $6p^4$	85 **At** $4f^{14}6s^25d^{10}$ $6p^5$	86 **Rn** $4f^{14}6s^25d^{10}$ $6p^6$
87 **Fr** $7s^1$	88 **Ra** $7s^2$	103 **Lr** $5f^{14}7s^26d^1$	104 **Rf** $5f^{14}7s^26d^2$	105 **Db** $5f^{14}7s^26d^3$	106 **Sg** $5f^{14}7s^26d^4$	107 **Bh** $5f^{14}7s^26d^5$	108 **Hs** $5f^{14}7s^26d^6$	109 **Mt** $5f^{14}7s^26d^7$	110	111	112		114		116		118
Lanthanide elements		57 **La** $6s^25d^1$	58 **Ce** $4f^26s^2$	59 **Pr** $4f^36s^2$	60 **Nd** $4f^46s^2$	61 **Pm** $4f^56s^2$	62 **Sm** $4f^66s^2$	63 **Eu** $4f^76s^2$	64 **Gd** $4f^75d^16s^2$	65 **Tb** $4f^96s^2$	66 **Dy** $4f^{10}6s^2$	67 **Ho** $4f^{11}6s^2$	68 **Er** $4f^{12}6s^2$	69 **Tm** $4f^{13}6s^2$	70 **Yb** $4f^{14}6s^2$		
Actinide elements		89 **Ac** $7s^26d^1$	90 **Th** $7s^26d^2$	91 **Pa** $5f^27s^26d^1$	92 **U** $5f^37s^26d^1$	93 **Np** $5f^47s^26d^1$	94 **Pu** $5f^67s^2$	95 **Am** $5f^77s^2$	96 **Cm** $5f^77s^26d^1$	97 **Bk** $5f^97s^2$	98 **Cf** $5f^{10}7s^2$	99 **Es** $5f^{11}7s^2$	100 **Fm** $5f^{12}7s^2$	101 **Md** $5f^{13}7s^2$	102 **No** $5f^{14}7s^2$		

Figure 3.4 The Periodic Table showing the atomic numbers and electronic configurations of the elements; those that are numbered, but unnamed, have been synthesized in small quantities

vention. The quantum rules define the different types of atomic orbitals which may be used for electron occupation in atoms. The Pauli exclusion principle defines the number of each type of orbital and limits each orbital to a maximum electron occupancy of two. Experimental observation, together with some sophisticated calculations, indicates the energies of the available orbitals for any particular atom. The **electronic configuration** (the orbitals which are used to accommodate the appropriate number of electrons) may be decided by the application of what is known as the ***aufbau*** (German: building-up) **principle**, which is that electrons in the ground state of an atom occupy the orbitals of lowest energy such that the total electronic energy is minimized.

3.2.1 The Electronic Configurations of the First Ten Elements

In the ground state of the hydrogen atom, the electronic configuration is that in which the electron occupies the 1s orbital, written as $1s^1$. The number of electrons occupying the orbital is indicated by the superscript.

For element number two (helium), there are two possible configurations which could be considered: $1s^2$ and $1s^12s^1$. There are two factors which decide which of the two configurations is of lower energy. These are the difference in energy between the 2s and 1s orbitals, and the greater interelectronic repulsion energy in the $1s^2$ case. If the 2s–1s energy gap is larger than the interelectronic repulsion energy between the two electrons in the 1s orbital, the two electrons will pair up in the lower orbital. The point may be made by calculations which make use of equation (2.3), repeated and modified here to include the Rydberg constant (given by equation 2.5):

$$E_n = -\frac{N_A \mu Z^2 e^4}{8\varepsilon_0^2 h^2}\left[\frac{1}{n^2}\right] = -N_A RchZ^2\left[\frac{1}{n^2}\right] \quad (3.8)$$

Equation (3.8) may be modified by including a term which is equal to the ionization energy of the hydrogen atom. The **ionization energy** of the hydrogen atom from its **ground state** (the lowest energy electronic state):

$$H(g) \rightarrow H^+(g) + e^- \quad (3.9)$$

may be calculated from equation (3.8). The process of ionization is equivalent to the electronic transition between the energy levels corresponding to $n = 1$ and $n = \infty$ (because the reference energy is taken to be zero when $n = \infty$). The energy required to ionize the electron from the hydrogen atom ($Z = 1$) is given by:

$$I_H = E_\infty - E_1 = 0 + N_A Rch/1^2 = N_A Rch \quad (3.10)$$

Worked Problem 3.3

Q Use equation (3.10) to calculate a value for the ionization energy of the hydrogen atom.

A The ionization energy of the hydrogen atom is given by the product:

6.0221367×10^{23} (mol^{-1}) $\times$ 10967760 (m^{-1}) $\times$ 299792458 (m s^{-1}) $\times$ $6.6260755 \times 10^{-34}$ (J s) = 1312 kJ mol^{-1}

[Note that the energy is identical to the photon energy emitted in the first Lyman line of the hydrogen spectrum; can you see why this should be?]

Making the substitution of I_H for $N_A Rch$ in equation (3.8) produces the equation:

$$E_n = -Z^2 I_H \left[\frac{1}{n^2}\right] \quad (3.11)$$

Worked Problem 3.4

Q Calculate a value for the ionization energy of the He^+ ion, assuming it to have its single electron in the lowest energy atomic orbital with a value of $n = 1$.

A Using equation (3.11):

$E_1 = -2^2 \times 1312 = -5248$ kJ mol^{-1}, giving a value for the ionization energy of He^+ of +5248 kJ mol^{-1}.

This equation is applicable to hydrogen and hydrogen-like atoms (*i.e.* those possessing only one electron, *e.g.* He^+, Li^{2+}, ...). The energy of the 1s orbital in the helium atom is given by putting $Z = 2$ and $n = 1$ in equation (3.11), giving $E(1s_{He}) = -4I_H$. The energy of the 2s orbital of helium is given by putting $n = 2$, so that $E(2s_{He}) = -I_H$. Ignoring the effect of interelectronic repulsion, the energy of the $1s^2$ configuration is

$2 \times -4I_H = -8I_H$, and is thus lower than that of the $1s^1 2s^1$ configuration of $-5I_H$. An estimate of the interelectronic repulsion energy in the $1s^2$ configuration allows a conclusion to be made as to which of the two possible configurations is appropriate to the ground state of the helium atom. The first ionization energy of the helium atom, *i.e.* the energy required to cause the ionization:

$$He(g) \rightarrow He^+(g) + e^-(g) \tag{3.12}$$

is observed experimentally to be 2370 kJ mol^{-1} (rather than $4I_H$ which is 5248 kJ mol^{-1}) and the second ionization energy, *i.e.* that needed to cause the change:

$$He^+(g) \rightarrow He^{2+}(g) + e^-(g) \tag{3.13}$$

is observed to be 5248 kJ mol^{-1} (*i.e.* exactly equal to $4I_H$). The discrepancy between the calculated and observed values of the first ionization energy of He gives an estimate of the interelectronic repulsion energy. The difference between the hydrogen-like calculated value and the observed value for the first ionization energy of the helium atom gives an estimate of the magnitude of the interelectronic repulsion energy of 5248 – 2370 = 2878 kJ mol^{-1}, an amount which is less than the 2s–1s energy gap, which is given by $3I_H$ (3936 kJ mol^{-1}). This ignores the interelectronic repulsion between the 1s and 2s electrons which would increase the gap, and the 1s–2s exchange energy (explained in Section 3.3.2) which would decrease it slightly if the electron spins were parallel.

Figure 3.5 shows the calculated and experimental energies which are used in determining the ground state configuration of the helium atom. There is, therefore, no doubt that the ground state of the helium atom has the configuration $1s^2$. **Electrons only pair up in the same orbital when it is the lowest energy option.**

Box 3.2 Nuclear Shielding in the Helium Atom

To apply the nuclear shielding concept to the helium ionization energies it is necessary to modify equation (3.11) to take into account the effective nuclear charge, Z_{eff}:

$$E_n = -Z_{eff}^2 I_H \left[\frac{1}{n^2} \right] \tag{3.14}$$

This equation now represents the lowest electronic level in the helium atom, and $-E_n$ may be equated with the first ionization energy of 2370 kJ mol^{-1} to give a value for Z_{eff}:

$Z_{eff} = (2370/1312)^{1/2} = 1.344$, the actual nuclear charge of 2 units being reduced by $2 - 1.344 = 0.656$ units, which is the magnitude of the shielding constant, σ, in the equation:

$$Z_{eff} = Z - \sigma \tag{3.15}$$

which defines the shielding constant.

The above calculation allows equation (3.14) to reproduce the first ionization energy of the helium atom, but the unmodified equation (3.11) would then be used to predict exactly the second ionization energy.

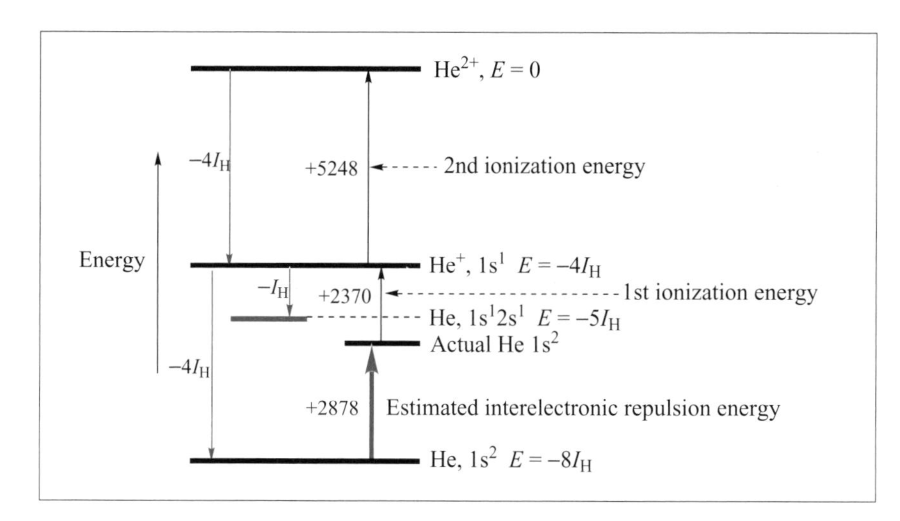

Figure 3.5 Energies of some He and He^+ configurations. The energy of the $1s^12s^1$ configuration of He does not include any interelectronic repulsion which would destabilize the system, and would make it quite certain that there is no question of the $1s^12s^1$ configuration representing the ground state of He

The *aufbau* order of filling of atomic orbitals (as the nuclear charge increases steadily along the order of the elements) can be written down in a pattern similar to the Periodic Table, as shown in Table 3.2. The filling begins on the top left-hand side with the 1s orbital, and then each successive row is filled, ending with the incompletely filled 7p orbitals. There are some irregularities in the filling which are dealt with later in this chapter, but the order of filling is generally in line with the value of the sum $n + l$; if there is more than one set of orbitals with the same sum, the set with the larger or largest l value is filled preferentially. For example, the 3d and 4p sets of orbitals have a sum of $n + l = 5$, but the 3d orbitals are filled before the 4p.

Notice the similarities between the two parts of Table 3.2 and the form of the Periodic Table given in Figure 3.4. The quantum rules, the Pauli exclusion principle and the *aufbau* principle combine to explain the general structure of the Periodic Table.

Table 3.2 An arrangement showing the relationship between the atomic orbitals filled and the number of elements in the various groups and periods of the Periodic Table. Both the major blocks mimic the arrangement of the elements in the 18-group Periodic Table

Orbitals filled				*Numbers of elements associated with the maximum filling of the orbitals*			
1s				2			
2s			2p	2			6
3s			3p	2			6
4s		3d	4p	2		10	6
5s		4d	5p	2		10	6
6s	4f	5d	6p	2	14	10	6
7s	5f	6d	7p[a]	2	14	10	6

[a] Incomplete at this time

The third element, lithium ($Z = 3$), has a full 1s orbital, with the third electron entering the 2s orbital to give the configuration $1s^22s^1$. Beryllium ($Z = 4$) has the configuration, $1s^22s^2$, the interelectronic repulsion energy in $2s^2$ being lower than the 2p–2s energy gap.

The fifth electron in the boron ($Z = 5$) atom enters one of the three-fold degenerate 2p orbitals. Since these are truly degenerate it is not proper to specify which of the three orbitals is singly occupied, although some texts choose the $2p_x$ orbital for alphabetical reasons. Boron has the electronic configuration $1s^22s^22p^1$. The carbon configuration is of considerable interest, since it is the first example of electrons occupying *degenerate orbitals*. From previous considerations it is a simple matter to conclude that the two electrons occupying the 2p orbitals should occupy separate orbitals, where interelectronic repulsion is less than if they doubly occupy a single orbital. What is not so obvious is that the two electrons, occupying two different 2p orbitals, should have identical values of the spin quantum number, m_s. The reason for this is that electrons with parallel spins (identical m_s values) have zero probability of occupying the same space: a kind of Pauli restriction, whereas two electrons with opposed spins (different m_s values) have a finite chance of occupying the same region of space within the atom and, in consequence, the interelectronic repulsion between them is greater than if their spins were parallel. This is the rationalization of **Hund's rules**. These may be stated in the following way:

"In filling a set of degenerate orbitals: (i) the number of unpaired electrons is maximized, and (ii) such unpaired electrons will possess parallel spins."

The electronic configuration of the carbon atom is $1s^22s^22p^2$ or, if the detailed content of the 2p orbitals is being discussed, it may be written as: $1s^22s^22p_x^{\ 1}2p_y^{\ 1}$, the choice of x and y being merely alphabetical. The 2p configuration is sometimes indicated diagrammatically by entering arrows (representing electrons with a particular spin) into two of three boxes (representing the 2p atomic orbitals in this case), as shown in Figure 3.6.

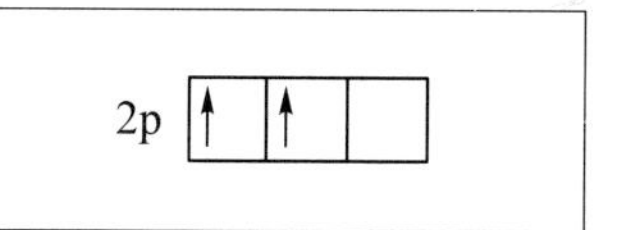

Figure 3.6 A "box/arrow" diagram indicating the spins of two unpaired electrons in the electronic configuration of the carbon atom; the direction of the arrows indicate the spin

Box 3.3 Box/arrow Diagrams

Box/arrow diagrams, boxes representing orbitals and arrows representing electrons with their spins, are only approximate descriptions of the electron arrangements in atoms. The possible arrangements of two electrons in a set of triply degenerate orbitals are usually described by diagrams such as the one shown in Figure 3.7.

The lowest diagram expresses the Hund's rules outcome, and the others represent the two excited states, one with electrons in separate orbitals to minimize interelectronic repulsion and the other with the electrons pairing up in one orbital with opposed spins.

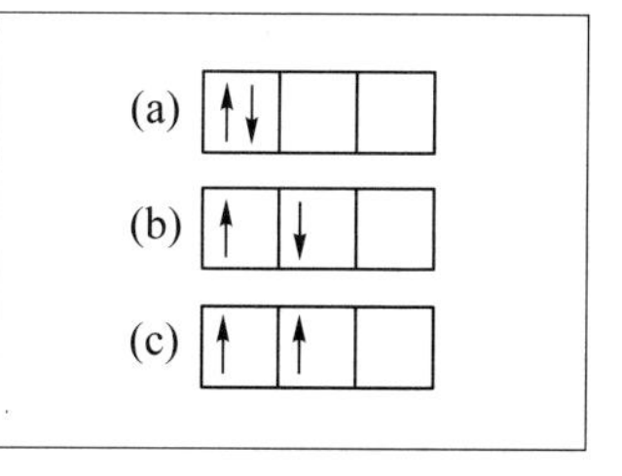

Figure 3.7 "Box/arrow" diagrams for possible arrangements of two electrons in a set of triply degenerate orbitals: (a) represents the highest energy state, which suffers maximum interelectronic repulsion, (b) represents the state which has minimum interelectronic repulsion, but has no exchange energy term (explained in Section 3.3.2), and (c) represents the ground state, which has both minimum interelectronic repulsion and exchange energy stabilization

By similar arguments it is concluded that the nitrogen atom has a lowest energy electronic configuration of $1s^22s^22p_x^{\ 1}2p_y^{\ 1}2p_z^{\ 1}$, the three 2p electrons having parallel spins. In the oxygen, fluorine and neon atoms the extra electrons doubly occupy the appropriate number of 2p orbitals, since pairing is the lowest energy option, the 3s–2p gap being greater than any interelectronic repulsion energy involved. The detailed electronic configurations of the elements of the second short period of the Periodic Table are given as "box" diagrams in Table 3.3.

Table 3.3 The electronic configurations of the first 10 elements[a]

Element	*1s*	*2s*	*$2p_x$*	*$2p_y$*	*$2p_z$*
H	↑				
He	↑↓				
Li	↑↓	↑			
Be	↑↓	↑↓			
B	↑↓	↑↓	↑		
C	↑↓	↑↓	↑	↑	
N	↑↓	↑↓	↑	↑	↑
O	↑↓	↑↓	↑↓	↑	↑
F	↑↓	↑↓	↑↓	↑↓	↑
Ne	↑↓	↑↓	↑↓	↑↓	↑↓

[a] The choice of filling the 2p orbitals alphabetically is arbitrary as they are three-fold degenerate.

3.3 The Electronic Configurations of Elements Beyond Neon

The atoms of the elements Li, Be, B, C, N, O, F and Ne form the second period of the Periodic Table. The third period contains the elements Na, Mg, Al, Si, P, S, Cl and Ar, which have **core electronic configurations** which are that of neon ($1s^22s^22p^6$) plus those derived from the regular filling of the 3s and 3p orbitals, as described above for the 2s and 2p orbitals. Identical arguments apply, and the elements of the third period are arranged under their counterparts in the second period with identical "outer electronic configurations", except for the change in value of n from 2 to 3. The term "outer" is a reference to the value of n, and is related to the greater diffuseness of orbitals as n increases in value: the orbitals become larger. It is *the nature of the outer electronic configuration which determines the chemistry of an element*, and it is for this reason that more emphasis is placed upon it rather than the arrangement of *all* the electrons of an atom.

3.3.1 The First Long Period

The next elements are potassium and calcium, in which the outer electrons occupy the next lowest energy orbital, which is the 4s since it has a lower energy than the 3d orbitals in the two cases. In the next 10 elements (scandium to zinc) the five 3d orbitals are progressively occupied. The filling of the 3d orbitals is in accordance with Hund's rules, with two irregularities which are described below.

The outer electronic configurations of the elements Sc, Ti and V are Sc $4s^23d^1$, Ti $4s^23d^2$, V $4s^23d^3$, but that of chromium is Cr $4s^13d^5$. Regularity returns with the next four elements: Mn $4s^23d^5$, Fe $4s^23d^6$, Co $4s^23d^7$, Ni $4s^2\,3d^8$. With copper, the 3d level is sufficiently lower than the 4s to ensure complete pairing up in the 3d orbitals, leaving the sole unpaired electron in the 4s orbital: Cu $4s^13d^{10}$. The first series of transition elements is completed by zinc, which has the configuration $4s^23d^{10}$.

Worked Problem 3.5

Q Write down the values of the four quantum numbers of the electrons that may occupy 3d orbitals.

A The 3d electrons all have $n = 3$ and $l = 2$. The permitted combinations of the values of m_l and m_s are:

m_l	m_s
2	$\pm\frac{1}{2}$
1	$\pm\frac{1}{2}$
0	$\pm\frac{1}{2}$
–1	$\pm\frac{1}{2}$
–2	$\pm\frac{1}{2}$

Thus, there is a maximum of ten 3d electrons in an atom of the third long period.

Worked Problem 3.6

Q Why are there ten groups of transition elements in the Periodic Table?

A Because the conclusions of the previous problem apply to any period with an *n* value of 3 or more.

3.3.2 Irregularity in the Electronic Configurations of the First Long Period

The filling of the 3d orbitals in the first long period is irregular at Cr and Cu. The latter is easily explained as the 3d level is well below the 4s at that stage and is filled first, and when it is full (as in Cu) the final electron enters the next lowest level, which is the 4s. The oddity that requires further explanation is the persistence of the $4s^2$ configuration in the elements from scandium to cobalt, with the exception of chromium with its $4s^1 3d^5$ configuration. That it is an oddity is because the energy of the 3d level is below that of the 4s. The explanation is because of the difference between energies of *particular occupied orbitals* and the *total energy of the configuration.*

Assuming that the core energy of the [argon] configuration remains constant, it is possible to differentiate between the configurations $[Ar]4s^2 3d^1$, $[Ar]4s^1 3d^2$ and $[Ar]3d^3$ for the scandium atom by considering the effects of (i) orbital energies, (ii) interelectronic repulsions and (iii) exchange energies. Ignoring the interaction between the valence electrons and the argon core, the orbital energies are given by:

$$E(4s^2 3d^1) = 2E_{4s} + E_{3d} \quad (3.16)$$

$$E(4s^1 3d^2) = E_{4s} + 2E_{3d} \quad (3.17)$$

$$E(3d^3) = 3E_{3d} \quad (3.18)$$

If, as is the case with the first transition series of elements, the 3d level is below that of the 4s, it would be expected that the $3d^3$ configuration would be the most stable, making the value of $3E_{3d}$ the most negative (remember that orbital energies are negative quantities with respect to the ionization level).

The effect of **interelectronic repulsion** must now be considered. This occurs between all pairs of electrons, and the ones contributing to the energies of the three scandium considerations are *added* to the three equations (3.16)–(3.18) to modify them to read:

$$E(4s^2 3d^1) = 2E_{4s} + E_{3d} + J_{ss} + 2J_{sd} \quad (3.19)$$

$$E(4s^1 3d^2) = E_{4s} + 2E_{3d} + 2J_{sd} + J_{dd} \quad (3.20)$$

$$E(3d^3) = 3E_{3d} + 3J_{dd} \quad (3.21)$$

where the subscripts on the J values indicate the types of electrons repelling each other.

The third factor to be considered is that of **exchange energy**. This stabilizing effect arises whenever two or more electrons with the same spin exist in a system. The interelectronic repulsion between two electrons with the same spin is less than that between two electrons with opposed spins. This is because, on average, electrons with parallel spins cannot approach each other as closely as can electrons with opposed spins. The effect is a kind of extension to the Pauli exclusion principle. The exchange energy (the stabilization of two electrons with the same spin) is denoted by the symbol K. The extent of exchange energy stabilization of a system depends upon the number of electrons with parallel spins.

Box 3.4 Calculation of the Number of Combinations of Electron Spins

For a system with n parallel spins, the stabilization is given by ${}^nC_2 \times K$, where ${}^nC_2$ represents the number of combinations of electron pairs: $n!/(2!(n-2)!)$ [the exclamation mark is used to indicate the **factorial** of a number, *e.g.* $4! = 4 \times 3 \times 2 \times 1 = 24$]. The $4s^1 3d^5$ configuration would have six electrons with parallel spins with an exchange energy of $15K$. The alternative configuration for Cr of $4s^2 3d^4$ would be higher in energy by the energy difference between the 4s and 3d levels, and would be less stable also by there being only five electrons with parallel spins, leading to a loss of exchange energy stabilization of $5K$. The same method may be used to calculate the number of J (Coulombic) interactions which are independent of the spin states of the electrons concerned.

Consideration of the exchange energy contributions between electrons of like spins modifies equations (3.19)–(3.21) to read:

$$E(4s^2 3d^1) = 2E_{4s} + E_{3d} + J_{ss} + 2J_{sd} - K_{sd} \quad (3.22)$$

$$E(4s^1 3d^2) = E_{4s} + 2E_{3d} + 2J_{sd} + J_{dd} - 2K_{sd} - K_{dd} \quad (3.23)$$

$$E(3d^3) = 3E_{3d} + 3J_{dd} - 3K_{dd} \quad (3.24)$$

where the subscripts of the K values indicate the types of electrons undergoing exchange. Electron exchange leads to an increase in stability; hence the K terms are given a *negative* sign. Equations (3.22)–(3.24) are sufficiently accurate for a decision to be made about which configuration is the most stable. Calculation of the individual J and K values is beyond the scope of this book, but because the 3d orbitals are more stable, electrons in them are more tightly bound than are electrons in the 4s orbital and repel each other to a greater extent. This makes the J_{dd} term dominant, and leads to the rather unusual conclusion that the $[Ar]4s^2 3d^1$ configuration is the most stable for Sc. The interaction of the various factors in similar equations for the other first-row transition elements explain their electronic configurations satisfactorily. In the elements Ti, V, Mn, Fe, Co and Ni the $4s^2$ pairs of electrons allow the minimization of the d–d repulsions, and only in the chromium case does the exchange energy stabilization play a major role when the 4s and five 3d orbitals are singly occupied with all six electrons having their spins in parallel.

In these considerations of 4s/3d occupancy, it is important to keep in mind that the orbital energies themselves are dependent upon the effective nuclear charge operating for a particular electronic configuration, so that the removal by ionization of one electron affects the orbital energies of the remaining electrons. This means that the ionization of scandium, $4s^2 3d^1$, removes one of the 4s electrons, and alters the orbital energies of the 4s and 3d electrons such that the remaining 4s electron finds greater stability in the 3d level, and with the gain of one unit of exchange energy produces a Sc^+ ion with the configuration $3d^2$. The ionization of the elements of the first transition series to give their +2 ions removes the $4s^2$ pair (from those elements which possess this pair) to give a $3d^{n-2}$ configuration (where n represents the total number of valence electrons, *i.e.* 4s + 3d in the individual atom). The removal of the electrons alters the 4s–3d energy gap to ensure that there are no instances of 4s orbital occupancy in the ionized species.

Irregularities in the configurations of the elements of the second and third transition series, and of the lanthanides and actinides, are described below, and their origins and explanations can be described in terms similar to those used for the first transition series.

The next orbitals to be used in building up the elements are the 4p set, which are filled in a regular fashion in the elements gallium to argon, thus completing the fourth period (or first long period) of the Periodic Table. The first two elements (K and Ca) are arranged so that they come below Na and Mg and form parts of Groups 1 and 2, respectively. The first set of transition elements form the first members respectively of Groups 3 to 12, and the elements from Ga to Kr are placed under those from Al to Ar as members of Groups 13 to 18.

3.3.3 The Second Long Period

The filling of the 5s, 4d and 5p orbitals accounts for the elements of the second long period: Rb and Sr in the **s-block**, Y to Cd in the second transition series or **d-block**, and In to Xe in the **p-block**. As was the case with the 3d orbitals, the filling of the 4d set is not regular owing to the closeness of the energies of the 5s and 4d orbitals. The irregularities are not the same as in the first set of transition elements. The outer electronic configurations of the second series of transition elements are: Y $5s^2 4d^1$, Zr $5s^2 4d^2$, Nb $5s^1 4d^4$, Mo $5s^1 4d^5$, Tc $5s^1 4d^6$, Ru $5s^1 4d^7$, Rh $5s^1 4d^8$, Pd $4d^{10}$ (*i.e.* no 5s electrons), Ag $5s^1 4d^{10}$ and Cd $5s^2 4d^{10}$.

In Y and Zr the 5s orbital is doubly occupied to avoid or minimize d–d repulsions, but the influence of exchange energy terms shifts the balance in the elements Nb, Mo, Tc, Ru and Rh so that the 5s orbital is singly occupied. In the elements Pd, Ag and Cd the 4d energy is distinctly lower than that of the 5s orbital, so that complete pairing occurs in Pd and then the 5s filling follows in Ag and Cd.

3.3.4 The Third Long Period

The next orbital of lowest energy to be used is the 6s, whose filling accounts for the outer electronic configurations of Cs and Ba, and then come the 5d and 4f sets whose energies are very nearly identical, but vary with the nuclear charge. The electronic configurations of the next 15 elements are shown in Table 3.4.

The 5d orbital is singly occupied at the beginning, in the middle and at the end of the above series of elements, which is where it has very similar energy to the 4f set of orbitals. The preference for the single 5d

Table 3.4 Outer electronic configurations of elements 57–71

Element	*Symbol*	*6s*	*5d*	*4f*
Lanthanum	La	2	1	0
Cerium	Ce	2	0	2
Praseodymium	Pr	2	0	3
Neodymium	Nd	2	0	4
Promethium	Pm	2	0	5
Samarium	Sm	2	0	6
Europium	Eu	2	0	7
Gadolinium	Gd	2	1	7
Terbium	Tb	2	0	9
Dysprosium	Dy	2	0	10
Holmium	Ho	2	0	11
Erbium	Er	2	0	12
Thulium	Tm	2	0	13
Ytterbium	Yb	2	0	14
Lutetium	Lu	2	1	14

electron is connected with the minimization of f–f repulsions. For the other elements, the 4f energy is sufficiently lower than that of the 5d for the latter not to be used. In gadolinium, the $5d^1 4f^7$ arrangement is preferred to $4f^8$, as it maximizes the exchange energy. The elements La to Yb form the **f-block**, and are the 14 elements concerned with the filling of the 4f orbitals. Lutetium ($6s^2 5d^1 4f^{14}$) has a full set of 4f orbitals, and a single 5d electron, and is placed as the first member of the third transition series. Some versions of the Periodic Table have lanthanum as the first member of the third transition series, the lanthanide elements (cerium to lutetium) coming between lanthanum and the second member of the third transition series, Hf. Both lanthanum and lutetium have a single 5d electron, lutetium possessing 14 4f electrons ($4f^{14}$) as do the other members of the third transition series. The 15 elements lanthanum to lutetium have a common oxidation state of +3 in which they behave chemically in a very similar manner, making their separation very difficult.

The next nine elements from hafnium to mercury complete the third transition series, although there are irregularities of filling the 5d orbitals since the 5d and 6s energies are close together. The outer electronic configurations of these elements are: Hf $6s^2 5d^2$, Ta $6s^2 5d^3$, W $6s^2 5d^4$, Re $6s^2 5d^5$, Os $6s^2 5d^6$ and Ir $6s^2 5d^7$, a regular filling of the 5d orbitals whose energies are lower than that of the 6s orbital, but whose d–d repulsions are minimized by the preferential filling of the 6s orbital. A single 6s electron is preferred in the platinum atom, $6s^1 5d^9$, and the 5d energy becomes significantly lower than that of the 6s in the case of gold, $6s^1 5d^{10}$. The filling of the 6s orbital is completed in mercury: Hg $6s^2 5d^{10}$.

The regular filling of the 6p orbitals accounts for the outer electronic configurations of the elements from thallium to radon, thus completing the third long period with its integral f-block elements.

3.3.5 The Fourth (Incomplete) Long Period

The fourth long period is incomplete but, as far as it goes, mirrors the third long period. The first two elements, francium and radium, have outer electronic configurations $7s^1$ and $7s^2$, respectively. The energies of the next orbitals to be filled (6d and 5f) are similar and give rise to irregularities, dependent on which of them is the lower in energy and by how much, and on the interelectronic and exchange energies of each configuration. The accepted outer electronic configurations are: Ac $7s^26d^1$, Th $7s^26d^2$ ($7s^26d^15f^1$ according to some sources), Pa $7s^26d^15f^2$, U $7s^26d^15f^3$, and Np $7s^26d^15f^4$. The irregularities persist up to plutonium, $7s^25f^6$, and after that there is the presumption that there is a regular filling of the 5f orbitals, much like that of the 4f orbitals. The exception is the reappearance of a single 6d electron in the case of curium (exchange energy is maximized as for gadolinium in the lanthanide series). The transplutonium elements are intensely radioactive and have short lifetimes, so that detailed studies of their electronic configurations have not been carried out and the published ones are somewhat speculative.

As for the lanthanide elements, the actinides (actinium to nobelium) are placed as a set of 14 elements in the separate f block, with element 104, lawrencium, being the first member of the fourth transition series. Alternative versions of the Periodic Table have actinium as the first member of the fourth transition series, and have lawrencium as the last of the actinide elements in the series of 14 from thorium to lawrencium. Lawrencium, as does lutetium, possesses a full f set of 14 electrons with a single outer d electron. Some versions of the Periodic Table dodge the above considerations and have 15 "lanthanide" elements, La–Lu, and 15 "actinide" elements, Ac–Lr, placed in the Group 3 column. The fourth transition series is unfinished, and the element with the highest Z value which has so far been synthesized is number 112. The names of the translawrencium elements (*i.e.* those with atomic numbers greater than 103) were settled in 1997 after many years of international argument. They have now been ratified by the International Union of Pure and Applied Chemistry (IUPAC), and are as shown in Table 3.5.

The syntheses of isotopes of elements 110, 111, 112, 114, 116 and 118 have been reported, but there are doubts about their verification.

3.4 The Periodic Table Summarized

The Periodic Table consists of 18 groups, corresponding to the filling of

Table 3.5 Names and symbols of the translawrencium elements

Atomic number	*IUPAC Name*	*Symbol*
104	Rutherfordium	Rf
105	Dubnium	Db
106	Seaborgium	Rf
107	Bohrium	Bh
108	Hassium	Hs
109	Meitnerium	Mt

the ns, $(n-1)$d [for values of $n \geq 3$] and np orbitals for the values of n up to seven, with the f-block elements, associated with the filling of the 4f and 5f sets of orbitals, situated separately.

There are three short periods for $n = 1$–3, since there are no available d orbitals at that stage. Hydrogen and helium form the first short period, with hydrogen placed as the first element in group one and helium as the first element in Group 18. The elements from Li to Ne and Na to Ar form the second and third short periods, respectively, although some older accounts describe them as the first and second short periods, discounting H and He. The fourth, fifth and sixth periods all contain 18 elements [associated with the filling of the ns, $(n-1)$d and np orbitals for $n = 4$, 5 and 6]. The sixth period also contains 14 lanthanide elements associated with the filling of the seven 4f orbitals. The seventh period is similar to the sixth in having an extra 14 elements (associated with the filling of the seven 5f orbitals), but some of the later elements have yet to be synthesized.

The majority of elements in any particular group have identical outer electronic configurations (apart from having varying values of the principal quantum number, n). As already described, it is not possible for there to be a completely regular filling of orbitals which would allow all members of a group to have exactly the same outer electronic configuration. However, this is usually not the case for any particular oxidation state of the elements of one group, which do tend to have identical outer electronic configurations. For example, the outer configurations of the elements Ni, Pd and Pt are $4s^2 3d^8$, $4d^{10}$ and $6s^1 5d^9$, respectively, but their +2 ions have the common d^8 configuration.

Summary of Key Points

1. The importance of interelectronic repulsion in the understanding of polyatomic systems was discussed, followed by discussions of orbital penetration and screening effects that decide the breakdown in degeneracy of the hydrogen-like orbitals.

2. The spin quantum number was introduced, leading to a discussion of the Pauli exclusion principle and the anti-symmetric properties of real wave functions.

3. The fundamental theoretical principles were used to describe the electronic configurations of the elements and the construct the general form of the modern Periodic Table.

4. Hund's rules were stated and rationalized, and the *aufbau* building-up principle was used to describe electronic configurations of the elements. Exceptions to regular fillings of sets of orbitals were discussed.

Problems

3.1. Use equation (3.11) to calculate the ionization energies of the hydrogen-like ions He^{+}, Li^{2+} and Be^{3+}.

3.2. Explain why the titanium atom has the electronic configuration $[Ar]4s^2 3d^2$, although the 3d level is at a lower energy than the 4s.

3.3. Write down the detailed electronic configurations of the elements of the third period as given by Hund's rules.

3.4. Write down the detailed electronic configurations of the elements V, Cr, Mn and Fe in addition to their argon cores.

3.5. The electronic configuration of the europium atom is $[Xe]6s^2 4f^7$. Explain how the seven 4f electrons are distributed.

3.6. The properties of the elements of the second period are atypical of their "group chemistry". The properties of the elements of the first transition series are atypical of their "group chemistry". The properties of the lanthanides are significantly different from those of the actinides. Give an explanation of these anomalies in terms of the appropriate valence orbitals.

Further Reading

G. Herzberg, *Atomic Spectra and Atomic Structure*, 2nd edn., Dover, New York, 1944.

D. O. Hayward, *Quantum Mechanics for Chemists*, Royal Society of Chemistry, Cambridge, 2002. A companion volume in this series.

4

Periodicity I: Some Atomic Properties; Relativistic Effects

This chapter deals with the variations, across the periods and down the groups of the Periodic Table, of (i) the ionization energies, (ii) electron attachment energies (alternatively known as electron gain energies and electron affinities), (iii) atomic sizes and (iv) electronegativity coefficients of the elements. The influence of relativistic effects is introduced.

Aims

By the end of this chapter you should understand:

- The periodicity of the first ionization energies of the elements
- The variations in successive ionizations energies
- The periodicity of the electron attachment energies of the elements
- The definitions of atomic size
- Variations in atomic sizes across periods and down groups
- Variations in the sizes of ions
- The definitions of electronegativity coefficients
- Variations of electronegativity coefficients across periods and down groups
- Relativistic effects on electron energies and atomic sizes

4.1 Periodicity of Ionization Energies

This section is mainly a discussion of the variations in the *first* ionization energies of the elements. For many elements, their *successive* ionization energies are known and these are referred to where appropriate.

The **first ionization energy** of an atom is the minimum energy required to convert one mole of the gaseous atom [in its ground (lowest energy) electronic state] into one mole of its gaseous unipositive ion:

$$A(g) \rightarrow A^{+}(g) + e^{-} \quad (4.1)$$

Its value is an indication of the effectiveness of the attraction of the nuclear charge on the electron which is most easily removed. The first ionization energies of the elements are plotted in Figure 4.1. There is a characteristic pattern of the values for the elements Li to Ne which is repeated for the elements Na to Ar, and which is repeated yet again for the elements K, Ca and Al to Kr (the s- and p-block elements of the fourth period). In the latter case the pattern is interrupted by the values for the 10 transition elements of the d-block. The fourth period pattern is repeated by the fifth period elements, and there is an additional section in the sixth period where the 14 lanthanide elements occur. In addition to this periodicity there is a general downward trend, and the ionization energy decreases down most of the groups. Both observations are broadly explicable in terms of the electronic configurations of the elements. Exceptions to the general trends are dealt with in Section 4.5.

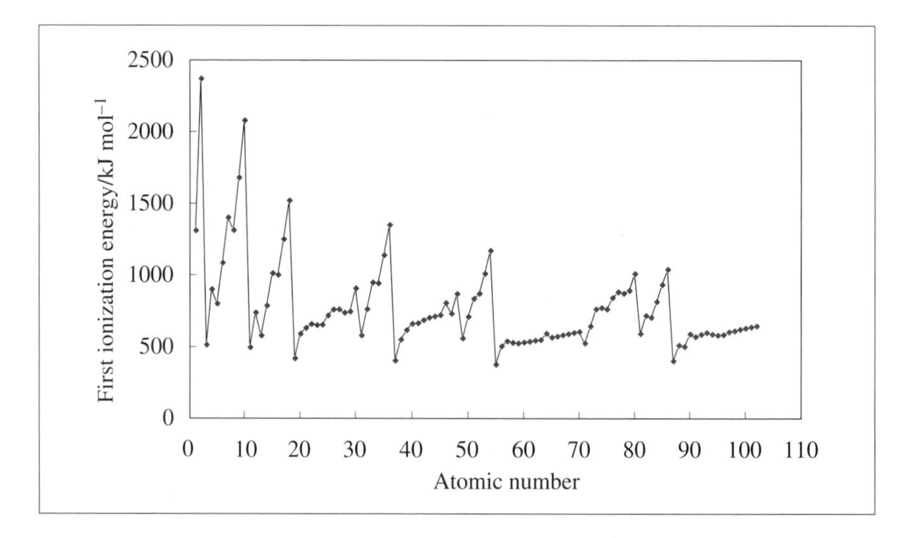

Figure 4.1 The first ionization energies of the elements

Box 4.1 Internal Energies and Enthalpy Changes

Atomic and molecular energy quantities, such as ionization energies, are tabulated in data books as values of **internal energy**, U, or changes in internal energy, ΔU, which are values at *constant volume and at 0 K*. A change in enthalpy, ΔH, is related to the change in internal energy by the equation: $\Delta H = \Delta U + P\Delta V = \Delta U + \Delta nRT$ (P is pressure, V is volume and Δn is the change in the number of moles of particles). In the case of ionization energies in which an atom is converted into a positive ion plus the liberated electron, $\Delta n = 1$. This means that the corresponding change in enthalpy, ΔH, is given by $\Delta U + RT$, and at the standard temperature of 298 K a further $\frac{3}{2}RT$ should be added to allow for the extra enthalpy possessed by the products of ionization at the temperature, T ($\frac{1}{2}RT$ for each of the extra degrees of translational freedom). This means that to convert an ionization energy to an ionization enthalpy an amount $\frac{5}{2}RT$ (6.2 kJ mol^{-1} at 298 K) should be added. Thus the ionization energy of the hydrogen atom, 1312 kJ mol^{-1}, is converted to the enthalpy of ionization by adding 6 kJ mol^{-1} to give 1318 kJ mol^{-1}. The difference is less than 0.5%, and is normally not of any great significance. In calculations in which atomic properties are mixed with true enthalpy changes, the difference should be acknowledged.

4.1.1 Variation of Ionization Energy Across a Period

The variation of the first ionization energies of the elements H and He, and those of the second period, Li–Ne, may be understood in terms of their electronic configurations and the changes in effective nuclear charge, together with differences in interelectronic repulsion energies and exchange energies. The electron which is removed in the ionization process is the one (or, in the case of *occupied degenerate orbitals*, is one of those) with the highest energy: the one requiring the least energy to remove from the attractive influence of the atomic nucleus. The terms used in the following discussion are the orbital energies, E_{1s}, E_{2s} and E_{2p} (all negative quantities related to the enthalpy of the ionized atom as the reference zero); the Coulombic repulsion energies between two electrons *doubly* occupying the 1s, 2s and 2p orbitals, J_{1s1s}, J_{2s2s} and J_{2p2p}; and the exchange energy, K. The repulsion energies between electrons in different orbitals are ignored in this discussion.

The ionization energy of the hydrogen atom is given by $-E_{1s}$. The increase in first ionization energy in moving from H to He is due to the increase in nuclear charge. As discussed in Section 3.2.1, the ionization energy of He is considerably less than that expected from the value of the nuclear charge because of the interelectronic repulsion between the two 1s electrons. Symbolically, the first ionization energy of helium is given by $-E_{1s} - J_{1s1s}$, the latter term being the Coulombic repulsion energy due to the double occupancy of the 1s orbital. Although E_{1s} for He is four times larger than it is for H (the influence of the Z^2 term in equation 3.11), the J_{1s1s} term causes the first ionization energy to be 2370 kJ mol^{-1}.

The changes in ionization energies along the second period are summarized in Table 4.1. The derivation of exchange energy contributions to the total energy is discussed in Chapter 3 and elaborated below.

Table 4.1 The first ionization energies of the elements Li to Ne and the theoretical contributions to their values

Atom	*First ionization energy/kJ mol⁻¹*	*Theoretical contributions*
Li	519	$-E_{2s}$
Be	900	$-E_{2s} - J_{2s2s}$
B	799	$-E_{2p}$
C	1090	$-E_{2p} + K$
N	1400	$-E_{2p} + 2K$
O	1310	$-E_{2p} - J_{2p2p}$
F	1681	$-E_{2p} - J_{2p2p} + K$
Ne	2081	$-E_{2p} - J_{2p2p} + 2K$

It should be noted that the orbital energies, E_{2s} and E_{2p}, are not constant along the period, but become increasingly negative in magnitude as the value of Z_{eff} increases. The combination of the three factors [varying orbital energies, exchange energy (K terms) and Coulombic energy (J terms) differences along the period] explain the observed pattern of changes in the first ionization energies.

A typical calculation of the ionization energy of nitrogen serves as a detailed example of the quantities given in Table 4.1. The electronic configuration of the nitrogen atom is $1s^22s^22p^3$, with the three 2p electrons following Hund's rules with their spins parallel. Considering the $1s^22s^2$ electrons as a "core" for these purposes, the energy of the nitrogen configuration is given by $E(\text{core}) + 3E_{2p} - 3K$. The removal of one electron gives the configuration $2p^2$, and the energy of the N^+ ion is given by $E(\text{core}) + 2E_{2p} - K$. The repulsion energies between the unpaired 2p electrons are ignored. The difference between the two energies gives the first ionization energy of the nitrogen atom as $-E_{2p} + 2K$.

Worked Problem 4.1

Q By doing calculations like the one above, check the entries in Table 4.1 for oxygen and fluorine atoms.

A Energy of O ($2p^4$) $= E(\text{core}) + 4E_{2p} + J_{2p2p} - 3K$

Energy of O^+ ($2p^3$) $= E(\text{core}) + 3E_{2p} - 3K$

Ionization energy $= (E(\text{core}) + 3E_{2p} - 3K) - (E(\text{core}) + 4E_{2p} + J_{2p2p} - 3K)$

$= -E_{2p} - J_{2p2p}$

the last term contributing to the reason for the first ionization energy of oxygen being lower than that for N. The other contributory factor is that there is a loss of two units of exchange energy in the nitrogen case, but no change in exchange energy in the oxygen ionization.

$$\text{Energy of F } (2p^5) = E(\text{core}) + 5E_{2p} + 2J_{2p2p} - 4K$$
$$\text{Energy of F}^+ (2p^4) = E(\text{core}) + 4E_{2p} + J_{2p2p} - 3K$$
$$\text{Ionization energy} = (E(\text{core}) + 4E_{2p} + J_{2p2p} - 3K) - (E(\text{core}) + 5E_{2p} + 2J_{2p2p} - 4K)$$
$$= -E_{2p} - J_{2p2p} + K$$

The value of the first ionization energy of the lithium atom is 519 kJ mol^{-1}, and corresponds to the change in electronic configuration $1s^22s^1$ to $1s^2$. The large decrease in moving from He to Li is because of the change in principal quantum number of the electron removed. The 2s electron of the lithium atom is influenced much less by the nuclear charge because of the shielding offered by the $1s^2$ core.

The increase to 900 kJ mol^{-1} in the case of the beryllium atom is due to the increase in effective nuclear charge, offset by interelectronic repulsion of the two 2s electrons. The electron most easily removed is one of the pair in the 2s orbital. In the case of the boron atom, in spite of an increase in nuclear charge there is a decrease in the first ionization energy to 799 kJ mol^{-1}. This is because the electron removed is from a 2p orbital, which is higher in energy than the 2s level.

The next two ionization processes, from C and N, are accompanied by changes in exchange energy in addition to the changes in nuclear charge. Compared to boron, there are increases in the first ionization energies of the carbon (1090 kJ mol^{-1}) and nitrogen (1400 kJ mol^{-1}) atoms. The electrons are removed from the degenerate 2p orbitals and the increases are due to the increase in nuclear charge, there being one and two units of exchange energy stabilization to be overcome, respectively.

With oxygen there is a slight decrease to 1310 kJ mol^{-1} in spite of an increase in nuclear charge, there being no difference in exchange energy between the $2p^4$ and $2p^3$ configurations. In the cases of B, C and N, the electron removed in the ionization process is the sole resident of one of the 2p orbitals. In oxygen there are two 2p orbitals which are singly occupied and one which is doubly occupied. Consideration of the higher interelectronic repulsion associated with a pair of electrons in the same orbital (represented by the J_{2p2p} term in Table 4.1) leads to the conclusion that it is one of the paired-up electrons in the 2p level of oxygen which is most easily removed. Consideration of Hund's rules leads to the conclusion that all three 2p electrons in the resulting O^+ ion possess parallel spins. The interelectronic repulsion-assisted removal of paired-up 2p electrons applies to the cases of the next two elements, F and Ne, the increases in first ionization energy being due to the increases in nuclear charge coupled with appropriate changes in exchange energies.

The general pattern of the variation of the first ionization energies of the elements of the second short period is repeated for the respective elements of the s and p blocks of the subsequent periods of the Periodic Table, except for those in period 6 (Cs, Ba, Tl, Pb, Bi, Po, At and Xe). This is because of relativistic effects, and the explanation of the different pattern is dealt with in Section 4.5.

In any of these series there are two variable quantities: the nuclear charge and the number of electrons of the atoms considered. It is possible to eliminate the effect of the nuclear charge (but not the effects of changes in the *effectiveness* of the nuclear charge) by considering, for example, the *successive ionization energies* of the neon atom. Figure 4.2 is a plot of the first eight successive ionization energies of the neon atom and, as would be expected from a reduction in the number of electrons being attracted by the constant nuclear charge, the values exhibit a general increase as ionization progresses.

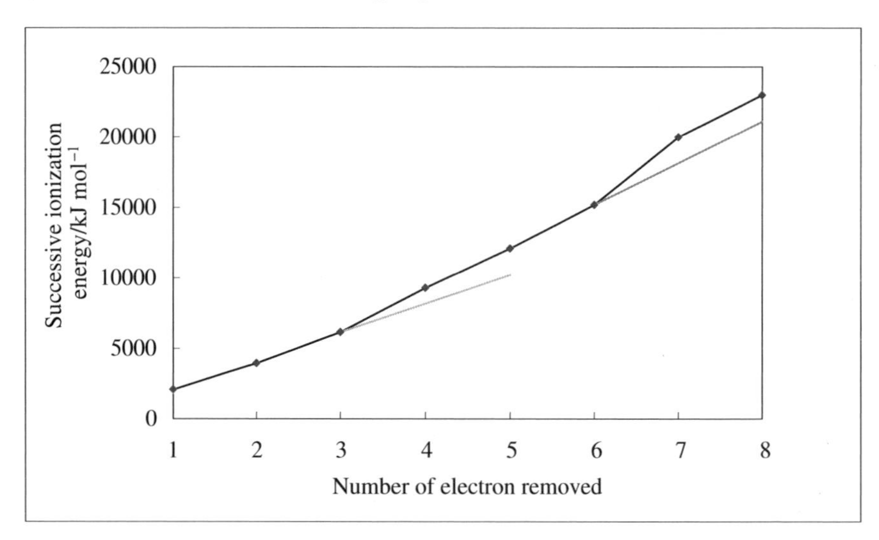

Figure 4.2 The eight successive ionization energies of the Ne atom

The first three ionizations arise from doubly occupied 2p orbitals, by the same reasoning as is given above for the first ionization of the oxygen atom. All three ionizations are assisted by the large interelectronic repulsion associated with the double occupation of orbitals (J_{2p2p}). As may be seen from the extrapolation of the line joining the first three points in Figure 4.2, the next three electrons are considerably more difficult to remove than the first three. There is an increasing effectiveness of the nuclear charge with the additional difficulty of electron removal owing to the absence of the high interelectronic-repulsion assistance experienced by the first three electrons to be removed.

The second discontinuity shown in Figure 4.2 is associated with the ionization of the seventh and eighth electrons, which are very much more difficult to remove because they originate in the 2s orbital of the neon atom.

The energy terms associated with the successive ionizations of the neon atom are given in Table 4.2 using the same approximations as those used to explain the pattern of first ionization energies for the atoms from Li to Ne. The values of E_{2p} and E_{2s} increase considerably in magnitude as Z_{eff}^2 increases to produce the observed variations.

Table 4.2 Contributions to the successive ionization energies of the neon atom

Electron removed	*Ionization energy*
First	$-E_{2p} - J_{2p2p} + 2K$
Second	$-E_{2p} - J_{2p2p} + K$
Third	$-E_{2p} - J_{2p2p}$
Fourth	$-E_{2p} + 2K$
Fifth	$-E_{2p} + K$
Sixth	$-E_{2p}$
Seventh	$-E_{2s} - J_{2s2s}$
Eighth	$-E_{2s}$

4.1.2 The Ionization Energies of the d-Block Elements

Because of the complications associated with the electronic configurations of the d-block elements owing to the closeness of energy of the 4s/3d, 5s/4d and 6s/5d sets of orbitals, the general trends of ionization energies of these elements are better indicated by their third ionizations. In such cases the corresponding s orbitals are vacant, and the effects of varying the nuclear charge and the number of d electrons may be studied. Figure 4.3 shows the variations of the third ionization energies of the three sets of transition elements: Sc to Zn, Y to Cd and Lu to Hg. There is the general increase expected as the effectiveness of the nuclear charge increases, with a discontinuity after Group 7 corresponding to ionization from a doubly occupied d orbital. The third ionizations from the atoms of Groups 8 to 12 are from doubly occupied d orbitals, and are consequently lower than would be expected if they did not have the repulsion assistance arising from two electrons in the same orbital.

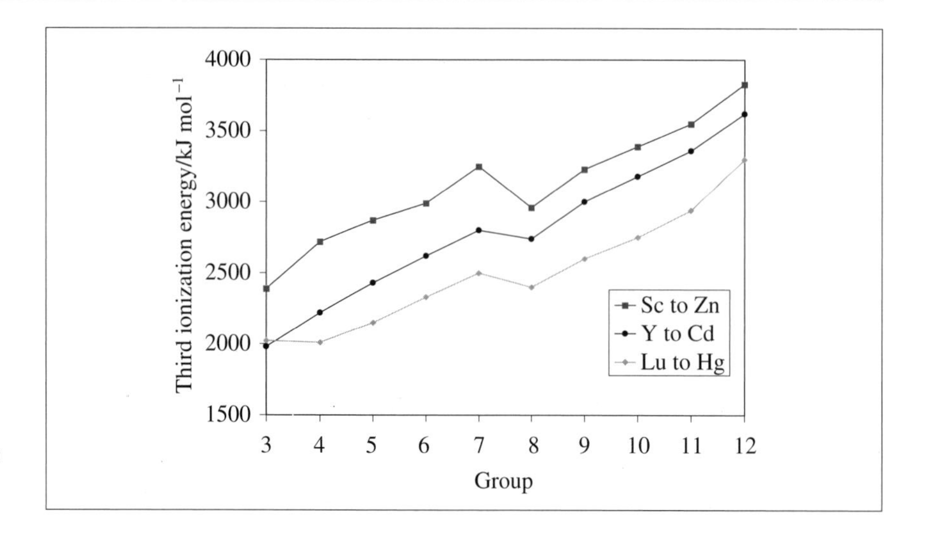

Figure 4.3 The third ionization energies of the transition elements

Worked Problem 4.2

Q Explain the discontinuity in Figure 4.3 in terms of orbital energies, interelectronic repulsion and exchange energies.

A The discontinuity occurs in the third ionization energies of Mn and Fe. Taking into account the interelectronic repulsion term for double occupancy of a 3d orbital, J_{3d3d}, and the exchange energy terms, K_{3d3d}, the energy of the d^5 configuration of Mn^{2+} is given by:

$$E(\mathrm{Mn}^{2+}) = 5E_{3d} - 10K_{3d3d}$$

and if one d electron is removed the energy of the resulting Mn^{3+} ion is given by:

$$E(\mathrm{Mn}^{2+}) = 4E_{3d} - 6K_{3d3d}$$

so the difference between these energies gives an expression for the third ionization energy of Mn as:

$$I_3(\mathrm{Mn}) = 4E_{3d} - 6K_{3d3d} - (5E_{3d} - 10K_{3d3d}) = -E_{3d} + 4K_{3d3d}$$

The loss of six units of exchange energy in going from the d^5 to d^4 configuration makes the third ionization energy particularly large. The calculation for the third ionization energy of Fe is:

$$E(\mathrm{Fe}^{2+}) = 6E_{3d} + J_{3d3d} - 10K_{3d3d}$$

the J_{3d3d} term expressing the interelectronic repulsion energy between the two paired-up electrons. If one d electron is removed, the energy of the resulting Fe^{3+} ion is given by:

$$E(Fe^{3+}) = 5E_{3d} - 10K_{3d3d}$$

so the difference between these energies gives an expression for the third ionization energy of Fe as:

$$I_3(Fe) = 5E_{3d} - 10K_{3d3d} - (6E_{3d} + J_{3d3d} - 10K_{3d3d}) = -E_{3d} - J_{3d3d}$$

There is no loss of exchange energy in this case, but the ionization energy is decreased by the amount of interelectronic repulsion.

4.1.3 Ionization Energies of the Lanthanide Elements

With the lanthanide elements, La to Lu, there is a less than regular filling of the 4f orbitals because of the proximity in energy of the 5d level. All the lanthanides possess a $6s^2$ pair of electrons and La, Ce, Gd and Lu have one 5d electron, so that the first ionization energies of the lanthanide elements are not easily interpreted. Figure 4.4 shows their third ionization energies, which exhibit a general increasing trend as the 4f orbitals are singly, then doubly, occupied according to Hund's rules. The large discontinuity going from Eu to Gd coincides with the change from the europium(III) $4f^7$ configuration, ionization from which is associated with the maximum loss of exchange energy, to the gadolinium(III) $4f^8$ configuration in which there is no loss of exchange energy, but the

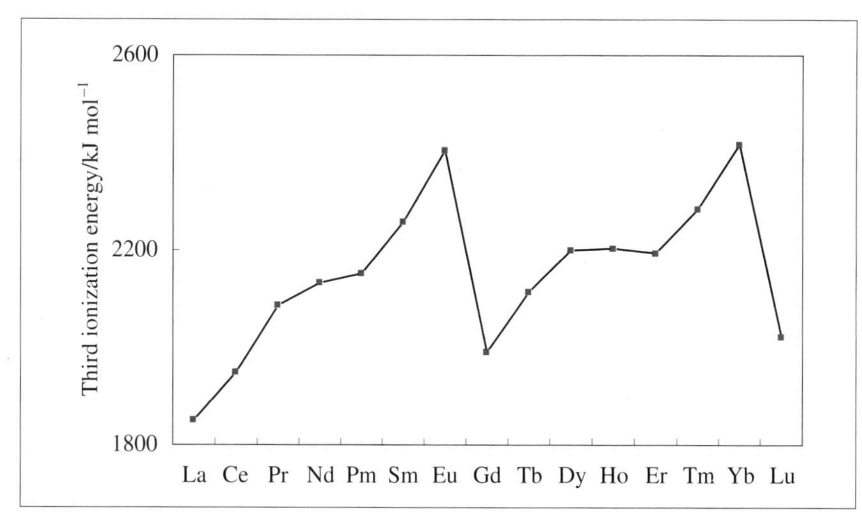

Figure 4.4 The third ionization energies of the lanthanide elements

interelectronic repulsion energy in the 4f orbital containing the pair of electrons causes ionization from the pair to be easier.

4.1.4 Variation of Ionization Energy Down a Group

In general there is a decrease in the first ionization energies of the atoms of any particular group of the Periodic Table as the nuclear charge increases. Down any group the orbital from which the electron is ionized has a progressively larger value of n, and has the highest energy for each particular atom in the group. This essentially offsets the effect of the increasing nuclear charge. It is as though the outermost electrons are *shielded* from the effect of the nuclear charge by the full inner sets of orbitals. A slight exception to the trend is noticeable in Group 13, in which the first ionization energy of gallium (579 kJ mol^{-1}) is marginally higher than that of aluminium (577 kJ mol^{-1}). This is because of the effect of the **3d contraction**, the general decrease in atomic size as the 3d orbitals are filled. The contraction causes increases in the ionization energies and this extends to the subsequent groups in the p-block. It is only obvious in the case of Al/Ga, but the first ionization energies of germanium, arsenic, selenium and bromine are all slightly higher than implied by their group trends.

The 3d contraction is discussed further in Section 4.3.2.

Further exceptions to the general trend in first ionization energies occur at the low ends of Groups 13 and 14, the first ionization energies of Tl and Pb being respectively higher than those of the corresponding elements in period 5, In and Sn. The exceptions are explained by relativistic effects that are dealt with in Section 4.5.

4.1.5 Successive Ionization Energies

The successive ionization energies of the neon atom are dealt with in Section 4.1.1 in some detail. Those of other atoms reveal similar trends and are subject to similar rationalizations. Additionally, the large increases in ionization energy associated with a change in the principal quantum number are of interest. Figure 4.5 is a graph of the eight successive ionization energies of the oxygen atom. It shows relatively small changes over the first six values, and there is then a large increase in the energy required to remove the seventh electron. This coincides with the seventh electron coming from the much more stable 1s orbital, rather than the 2p and 2s orbitals from where the first six electrons originate. The eighth ionization corresponds to the removal of the single remaining 1s electron in the O^{7+} ion. Since O^{7+} is hydrogen-like, the energy of the 1s orbital is derivable from equation (3.11), and there is complete agreement between theory and observation.

Successive ionization energies are useful in characterizing the differ-

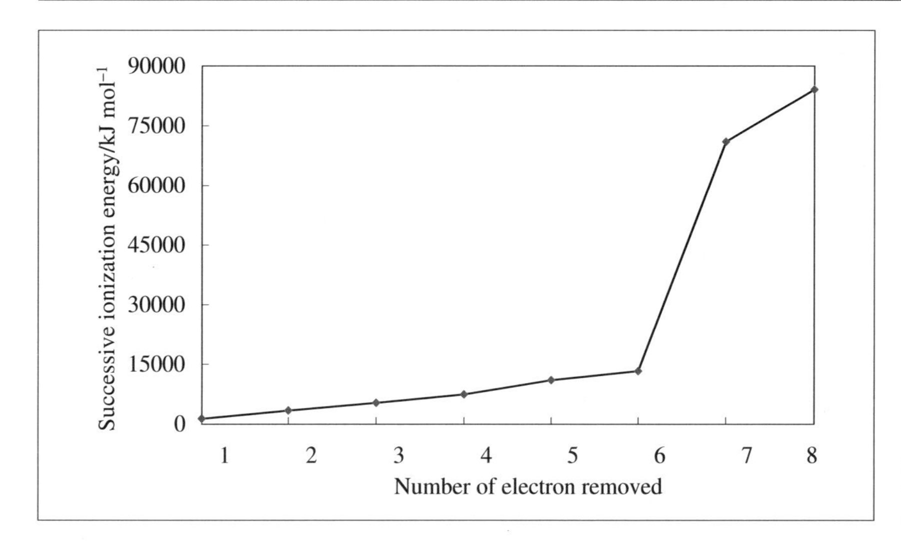

Figure 4.5 The eight successive ionization energies of the oxygen atom

ences in properties between the first member of a group of elements in the Periodic Table and the other members of that group. Taking Groups 13, 14 and 15 as examples, Figure 4.6 shows plots of the sums of the successive ionization energies of their valence electrons, the first three ionizations for Groups 13 and 15 and the first four for Group 14 (consistent with their usual valencies). It is clear that the electrons of the first members of each group are much more tightly bound than those of the remaining members of the groups. The values of the sums for the elements in Period 4 are all higher than might be expected from the group trends, and this is because of the 3d contraction (Section 4.3). There are contractions associated with the fillings of the 4d and 5d orbitals, but these have a much smaller effect than that associated with the filling of the 3d orbitals.

Examples of the glaring differences in properties between the first member of a group and the remaining elements of the group include: (i) boron is an electronegative non-metal but aluminium, gallium, indium and thallium are metals; (ii) the +4 oxide of carbon is the gaseous molecule CO_2 but the oxides of silicon, germanium, tin and lead are solid, giant arrays of the same formula EO_2; and (iii) nitrogen exists as the diatomic N_2 molecule, but phosphorus has allotropes which are the white form, consisting of discrete P_4 molecules, and the red (amorphous) and three black forms which are polymers of the P_4 unit.

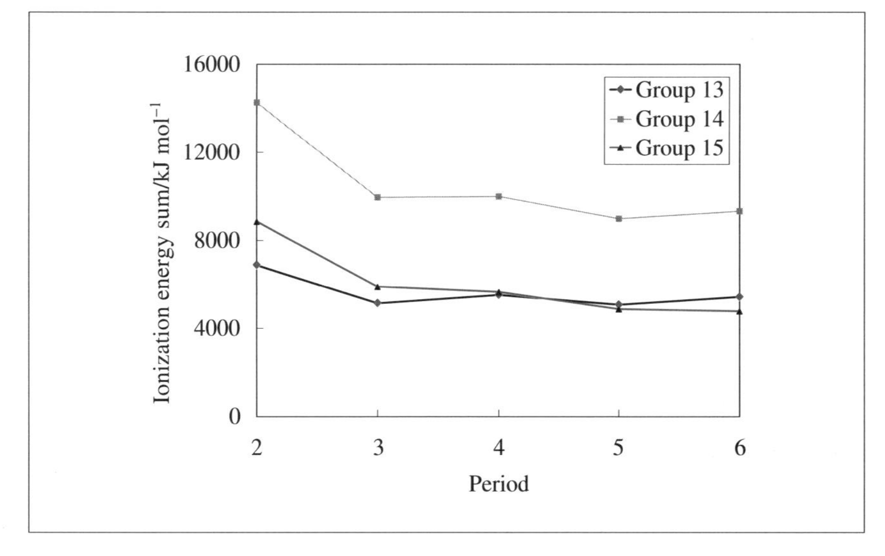

Figure 4.6 Plots of the sums of the successive ionization energies for the elements of Groups 13, 14 and 15. The sums for Groups 13 and 15 are of the first three ionization energies, those for the Group 14 elements are of the first four ionization energies

4.2 Variations in Electron Attachment Energies

The **electron attachment energy** or **electron gain energy** or **electron affinity** is defined as the *change* in internal energy (*i.e.* ΔU) that occurs when one mole of gaseous atoms of an element is converted by electron attachment to give one mole of gaseous negative ions:

$$A(g) + e^- \rightarrow A^-(g) \qquad (4.2)$$

1. Conversion to an electron attachment enthalpy requires the subtraction of $^5/_2RT$, i.e. 6.2 kJ mol^{-1} at 298 K. $\Delta n = -1$ for two particles (atom + electron) producing one particle (the negative ion).

2. In some older texts the electron affinities are quoted as positive quantities; in these cases the electron affinity is defined as the energy *released* when the negative ion is produced, and is of opposite sign to the electron attachment energies quoted in this and other modern texts.

Most elements are sufficiently electronegative to make their *first* electron attachment energies *negative* according to the thermodynamic convention that **exothermic reactions** are associated with **negative energy changes**. All second and subsequent electron attachment energies are *highly endothermic* because of the difficulty of adding a negative electron to an already negative ion.

The first electron attachment energies of the first 36 elements are plotted in Figure 4.7 and show the values for H and He followed by a characteristic pattern, the second repetition of which is split by the values for the 10 transition elements. The value for hydrogen is –72.8 kJ mol^{-1}, which is very different from the 1s orbital energy of –1312 kJ mol^{-1} because of the interelectronic repulsion term amounting to –72.8 – (–1312) = +1239.2 kJ mol^{-1}. The value for the helium atom is positive (+21 kJ mol^{-1}), indicating an overall repulsion of the incoming electron from the $1s^2$ configuration which is not offset by the nuclear charge. The incoming electron would occupy the higher energy 2s orbital.

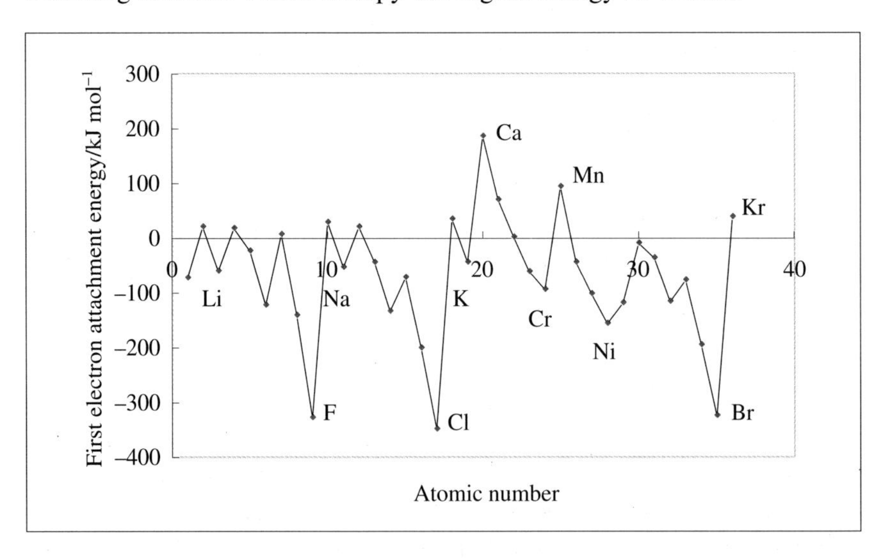

Figure 4.7 The electron attachment energies of the first 36 elements

The values of the electron attachment energies for the elements of the second short period (Li to Ne) may be understood by the variations in nuclear charge, and in interelectronic repulsion and exchange energies along the series. The changes are given in Table 4.3. The terms in Table 4.3 are the same as those used in Tables 4.1 and 4.2 and express the changes in orbital energy, interelectronic repulsion energies arising from the *double occupancy* of an atomic orbital (J_{2p2p}), and exchange energy (K) resulting from the electron attachment process.

For the carbon atom the $2p^2$ configuration has the energy: $E(\text{core}) + 2E_{2p} - K$; that of the C^- ion is: $E(\text{core}) + 3E_{2p} - 3K$; the difference gives $E_{2p} - 2K$ as the electron attachment energy for the carbon atom.

Table 4.3 Contributions to the first electron attachment energies of the elements Li to Ne

Element	*Electronic change*	*Electron attachment energy*
Li	$2s^1 \rightarrow 2s^2$	$E_{2s} + J_{2s2s}$
Be	$2s^2 \rightarrow 2s^22p^1$	E_{2p}
B	$2p^1 \rightarrow 2p^2$	$E_{2p} - K$
C	$2p^2 \rightarrow 2p^3$	$E_{2p} - 2K$
N	$2p^3 \rightarrow 2p^4$	$E_{2p} + J_{2p2p}$
O	$2p^4 \rightarrow 2p^5$	$E_{2p} + J_{2p2p} - K$
F	$2p^5 \rightarrow 2p^6$	$E_{2p} + J_{2p2p} - 2K$
Ne	$2p^6 \rightarrow 2p^63s^1$	E_{3s}

Worked Problem 4.3

Q Check the derivation of the terms in Table 4.3 for the nitrogen and oxygen atoms.

A Energy of N ($2p^3$) = $E(\text{core}) + 3E_{2p} - 3K$
Energy of N^- ($2p^4$) = $E(\text{core}) + 4E_{2p} + J_{2p2p} - 3K$
Electron attachment energy of N = $(E(\text{core}) + 4E_{2p} + J_{2p2p} - 3K) - (E(\text{core}) + 3E_{2p} - 3K) = E_{2p} + J_{2p2p}$

Energy of O ($2p^4$) = $E(\text{core}) + 4E_{2p} + J_{2p2p} - 3K$
Energy of O^- ($2p^5$) = $E(\text{core}) + 5E_{2p} + 2J_{2p2p} - 4K$
Electron attachment energy of O = $(E(\text{core}) + 5E_{2p} + 2J_{2p2p} - 4K) - (E(\text{core}) + 4E_{2p} + J_{2p2p} - 3K) = E_{2p} + J_{2p2p} - K$

The increase in electron attachment energy in moving from Li to Be is mainly because the added electron in Be has to occupy one of the otherwise vacant 2p orbitals, which is shielded from the effect of the nucleus by the two electron pairs $1s^22s^2$ of the Be atom. That shielding, or the general repulsion that the added electron experiences from the other four electrons, makes the electron attachment energy of Be slightly positive, *i.e.* the attachment process is *endothermic*. With boron, the added electron achieves some stability due to exchange with the existing electron in a singly occupied 2p orbital. The extra exchange energy given out when an electron is added to the carbon atom ensures a more negative value for the electron attachment energy in C. The increase on moving from C to N is due to the major amount of interelectronic repulsion energy when the added electron is forced to pair up with one of the existing 2p electrons. In oxygen and fluorine, the electron attachment energies become considerably negative, due to the amounts of exchange stabilization of the O^- and F^- ions, coupled with the general increase in the stability of the 2p atomic orbital as the nuclear charge increases along the series. With neon the added electron suffers general repulsion from the 10 electrons already present as it enters the relatively unstable 3s atomic orbital.

The same pattern as for Li to Ne is repeated for the elements Na to Ar and for the elements K, Ca and Ga to Kr, the latter series from the first long period being interrupted by the 10 transition elements. The variations along the 10 transition elements are caused by the general increase in the stability of the 3d orbitals as the nuclear charge increases, plus the effects of exchange energy changes and the enforced electron pairing in manganese and the elements that follow manganese. In the case of zinc, the added electron is forced to enter the 4p level. The remainder of the s- and p-block general pattern comes after zinc.

Worked Problem 4.4

Q Account for the first electron attachment energy of the calcium atom (+186 kJ mol^{-1}) being greater than that of the potassium atom (–48 kJ mol^{-1}).

A An electron attaching itself to a potassium atom would occupy the 4s orbital, pairing up with the valency electron already present. There would be interelectronic repulsion between the two 4s electrons, but this is overcome by the effective nuclear charge of the K^- ion. In the calcium case the 4s orbital is already full, and the extra electron must find space in a 3d orbital, which is the next level of lowest energy.

The electron attachment energies for the elements F and Cl are –328 and –349 kJ mol^{-1}, respectively. The usual reason advanced for the value for fluorine being less negative than that for chlorine is the higher repulsion between the more tightly bound lone pairs of electrons in F, compared to the more diffuse lone pairs of Cl (see the following worked problem).

Worked Problem 4.5

Q Consider a more quantitative way of expressing the differences between the electron attachment energies of F and Cl, using the J (for interelectronic repulsion) and K (for exchange energy) nomenclature, as is used for the discussion of trends in ionization energies and electron attachment energies in this section.

A Using the J_{2p2p} term that represents the *extra* interelectronic repulsion arising from the double occupancy of a 2p orbital, the electron attachment energy of the fluorine atom may be written as $E_{2p} + J_{2p2p} - 2K_{2p2p}$, as given in Table 4.3. The corresponding expression for the chlorine atom is $E_{3p} + J_{3p3p} - 2K_{3p3p}$. The value of E_{3p} for Cl is not as negative as the value of E_{2p} for F (their first ionization energies are F, 1681 and Cl, 1251 kJ mol^{-1}), indicating the greater stability of an electron in the 2p orbital of F than one in the 3p orbital of Cl. The dominant factors, therefore, in determining that the electron attachment energy of F is not as negative as that of Cl are the J_{2p2p} and J_{3p3p} terms. Because the 2p orbitals are closer to the nucleus and are less shielded than the 3p orbitals, the reason for the less negative value of the electron attachment energy of F is that J_{2p2p} for F must be considerably greater than J_{3p3p} for Cl.

4.2.1 Attachments of More than One Electron

Unlike the wealth of data available for successive ionization energies, there is very little data for the successive attachment of electrons to atoms. This is because of the repulsion between a uni-negative ion and a potential incoming electron. The second electron attachment energies of oxygen and sulfur are known: that of oxygen is +844 kJ mol^{-1} and that of sulfur is +532 kJ mol^{-1}. The Coulombic repulsion outweighs the gain in exchange energy as the second electron is added. That the additions of second electrons to oxygen and sulfur atoms are endothermic indicate that such processes are thermodynamically unfeasible, and they occur only in conjunction with the exothermic processes which accompany the formation of oxides and sulfides. The three successive electron attachment energies of the nitrogen atom are +7, +800 and +1290 kJ mol^{-1}, respectively.

4.3 Variations in Atomic Size

The size of an atom is not a simple concept. An inspection of the wave function for any atom shows that it is *asymptotic to infinity* (*i.e.* ψ becomes properly zero only at an infinite distance from the nucleus), so some practical definition of size is required. The previously discussed atomic properties are those of isolated gaseous atoms, but when sizes are discussed it is essential to consider the physical form of an element, whether it is monatomic like the Group 18 gases, whether it is metallic (in which case each atom is surrounded by between 12 and 14 nearest neighbours) or whether it is molecular and participating in covalent bonding.

All the atomic orbital wave functions contain the exponential term $e^{-\rho}$, where $\rho = 2Z_{eff}r/a_0$, which is zero only when $r = \infty$.

Box 4.2 Chemical Bonding

Chemical bonding may be summarized in terms of (i) **covalent bonding**, (ii) **ionic bonding** and (iii) **metallic bonding**.

This very brief account of bonding types is extended in Chapter 5, and dealt with fully in another book in this series (*Structure and Bonding*).

Covalent bonding

A covalent bond between two atoms is an **electron-pair bond** formed when both of the two participating atoms supply one electron, the electron pair being shared between the two atoms. This arrangement is a **single covalent bond**. Depending on their valency, some atoms form either more than one single bond or participate in multiple bonding (*i.e.* form double or triple bonds). **Coordinate** or **dative bonding** is a special type of covalency in which both electrons forming the electron-pair bond are supplied by one of the participating atoms. This form of bonding is particularly important in the **complex ions** formed by many transition elements, *e.g.* $[Co(NH_3)_6]^{3+}$ in which the ammonia molecules each supply a pair of electrons to form coordinate bonds to the central Co^{III} ion.

Ionic bonding

This only occurs between two different elements, one being electropositive the other being electronegative, when an electron is transferred from the electropositive element to the electronegative element to produce two ions which then are stabilized by electrostatic attraction in an **ionic lattice** or crystal. Many ions also are stable in solution, where they are stabilized by solvation (interaction between the ion and the solvent).

Metallic bonding

This occurs in elements which exist in the metallic state with each atom surrounded by more neighbours than it has electrons with which to form electron-pair bonds. The valency electrons occupy **bands** of orbitals which are **delocalized** throughout the metal and allow good electrical conduction to occur.

4.3.1 Estimates of Atomic Size

There are four different modes by which sizes may be assigned to any particular atom in its elementary or combined states.

Atomic Radius

The **atomic radius** of an element may be considered to be half the interatomic distance between two adjacent atoms. This may apply to iron, say, in its metallic state, in which case the quantity may be regarded as the **metallic radius** of the iron atom, or to a molecule such as Cl_2. In the latter case the quantity is the **covalent radius** of the chlorine atom. If applied to the multiply bonded O_2 and N_2 molecules, the radii would only be appropriate for oxygen and nitrogen atoms participating in multiple bonding in their compounds. The differences between these examples is sufficient to demonstrate that some degree of caution is necessary when comparing the atomic radii of different elements. It is best to limit such comparisons to elements with similar types of bonding, metals for example. Even that restriction is subject to the drawback that the metallic elements have at least three different **crystalline arrangements** with possibly different **coordination numbers**: the number of **nearest neighbours** for any one atom.

Covalent Radius

The covalent radius of an element may be considered to be one half of the covalent bond distance of a molecule such as Cl_2 (equal to its atomic radius in this case), where the atoms concerned are participating in single bonding. Covalent radii for participation in multiple bonding are also quoted in data books. In the case of a single bond between two different atoms, the bond distance is divided up between the participants by subtracting from it the covalent radius of one of the atoms, whose radius is known. A set of mutually consistent values is now generally accepted and, since the vast majority of the elements take part in some form of covalent bonding, the covalent radius is the best quantity to consider for the study of general trends. Only atoms of the Group 18 elements (except Kr and Xe) do not have covalent radii assigned to them, because of their general inertness with respect to the formation of molecules. The use of covalent radius for comparing the sizes of atoms is subject to the reservation that its magnitude, for any given atom, depends on the oxidation state of that element.

Ionic Radii

As indicated above, the covalent radius of an element depends on its oxidation state. In a binary ionic compound, MX, containing the positive ion, M^+, and the negative ion, X^-, the minimum distance between them is measurable with considerable accuracy by the method of X-ray diffraction. The problem is to divide such a distance into the ionic radii for the individual ions. That ions behave like hard spheres with a constant radius whatever their environment might be is an approximation to the real situation. In compounds which do not exhibit much covalency the approximation is reasonable, and led Shannon[3] and Prewitt to assign radii to O^{2-} and F^- of 140 and 133 pm respectively after their study of many oxides and fluorides. Ionic radii are not assignable to every element, and the generalizations described apply only to those elements which do form ions in compounds, and are subject to their oxidation states (discussed in Chapter 5) and coordination numbers (*i.e.* the number of nearest neighbours they have in the ionic compound).

Van der Waals Radius

The van der Waals radius of an element is half the distance between two atoms of an element which are as close to each other as is possible without being formally bonded by anything except van der Waals intermolecular forces. Such a quantity is used for the representation of the size of an atom with no chemical bonding tendencies: the Group 18 elements. That for krypton, for instance, is half of the distance between nearest neighbours in the solid crystalline state, and is equal to the atomic radius. Van der Waals radii of atoms and molecules are of importance in discussions of the liquid and solid states of molecular systems, and in the details of some molecular structures where two or more groups attached to the same atom may approach each other.

There are three types of van der Waals forces:

(i) interactions between molecules with permanent dipole moments; these are called dipole–dipole forces;

(ii) dipole–induced-dipole forces, where a molecule with a permanent dipole moment induces a dipole in a neighbouring molecule; and

(iii) London dispersion forces which operate between atoms.

The first two interactions are not relevant to elementary cohesion, but the third and weakest force is that which holds the Group 18 elements in their solid state. The London force arises because of a transitory and ever-changing deviation from spherical symmetry of an atom because the uncertainty principle forbids the accurate placement of the nucleus in the surrounding electronic configuration. The very small transient dipoles of atoms interact to give some cohesion to the Group 18 elements. Of course, the forces operate between all atoms, but if any kind of chemical bonding is present they represent an almost insignificant contribution to the overall cohesion.

Atomic Volume

The Periodic Law was first stated by Mendeleev[1] in 1869 as:

"The elements, if arranged according to their atomic weights, show an evident periodicity of properties."

In the same paper he published the first version of his Periodic Table. One year later, Lothar Meyer[2] supported Mendeleev's conclusions in a paper which had an essentially identical table. Meyer's paper contains the famous *atomic volume curve* showing maxima and minima as the atomic weight increased.

An up-to-date atomic volume plot is shown in Figure 4.8. The atomic volume is the volume of one mole of the element either in its standard state at 298 K or, if the element is not solid at this temperature, is the volume of one mole of the solid element below its melting point. The atomic volume of an element is easily calculated from its molar mass (RAM $\times$ 1 g mol^{-1}) and its density in the solid state:

$$\text{Atomic volume (cm}^3\text{ mol}^{-1}) = \text{RAM} \times (1\text{ g mol}^{-1}) \div \text{density (g cm}^{-3}) \quad (4.3)$$

and is plotted against the atomic number of the elements in Figure 4.8, rather than against atomic weight (RAM).

Figure 4.8 is a "boots and all" approach to the phenomenon of periodicity, but exemplifies the periodic law very well. Each period begins with a maximum value of atomic volume, goes through a minimum around half-way across a period, and then rises to the next maximum at the start of the next period. The cohesion of the solid elements, depending upon their individual characteristics, has varied contributions from metallic, covalent and van der Waals interactions, and so the atomic volume curve is best analysed by a wise choice of which parameter to plot against the atomic number. The atomic volume curve of Figure 4.8 can be understood in very general terms, with each period beginning with a loosely bonded metal belonging to Group 1. The weak metallic bonding,

The term relative atomic mass (RAM) has displaced atomic weight as an indication of the masses of atoms. Atomic weights, both in Mendeleev's time and until relatively recently, were estimated relative to the hydrogen atom having a value of exactly 1.0000.

Atomic number was only properly understood after the work of Geiger and Marsden, which showed that the nuclear charge in electron units was about half the "atomic weight" for the lighter elements. In 1911, Van den Broek of Utrecht suggested that the nuclear charge on an atom was equal to the ordinal number of that element in the Periodic Table; in 1913, Moseley of Manchester and Oxford supplied experimental proof that this was so.

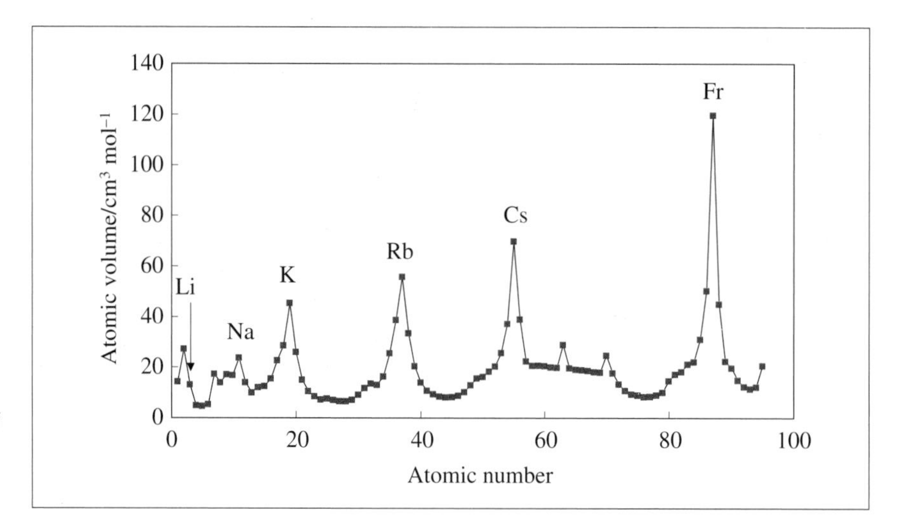

Figure 4.8 A modern version of the Lothar Meyer atomic volume curve: atomic volumes plotted against atomic numbers

because each atom has only one valence electron, causes the elements to have low densities and high atomic volumes. As the number of valency electrons increases across a period, the bonding becomes stronger and there is a transition from metallic to covalent bonding when there are sufficient electrons to produce electron-pair bonds. The minima of the plot coincide with the maximum usage of electrons in metallic bonding, and when electron shells (the d orbitals in particular) become more than half-full the bonding weakens again. Covalency becomes more important towards the end of each period, and the density is a function of the stronger covalent bonding, which draws the bonded atoms closer together, and the much weaker van der Waals intermolecular forces which are responsible for the cohesion of the molecular solids. This mixture of strong and weak interactions results in the rise in atomic volume at the end of each period, leading up to the next maximum at the beginning of the next period. The only slight exception to this generality is shown by lithium, which is not at a maximum in the curve. This is because of the exceptionally high atomic volume of helium. The London forces in helium are very slight and allow the helium atoms to have a particularly large radius and consequently the density of solid helium is very low. Down Group 18 the London forces increase considerably in their effectiveness, and no further exceptions arise.

4.3.2 Variations in Metallic Radii

Figure 4.9 shows how the metallic radii vary in Periods 2–6, excepting the lanthanide elements which are dealt with below. There is a general rapid reduction in metallic radius in each period followed by less rapid reduction, and finally there is a slight increase as the relevant d shell is eventually filled. These changes are consistent with there being more electrons to participate in the metallic bonding in the early portions of the periods, but this is offset more and more as electron pairing starts to predominate. Overall, there is a general increasing trend in the effective nuclear charge across a period, and superimposed on this are the electronic contributions to metallic bonding which are large and positive at first, but reduce in effectiveness towards the ends of the series shown. Down the groups there is a significant increase in metallic radius, particularly with Groups 1 and 2, but the sizes of members of the fifth and sixth periods are almost identical. It would be expected that the effectiveness of the nuclear charge would decrease down the groups as more and more electron shells contribute to the shielding of the valence electrons. This is so in general, but the coincidences in the fifth and sixth period need further explanation, which is given below. The size changes towards the end of each period are similar to those of the Meyer curve shown in Figure 4.8.

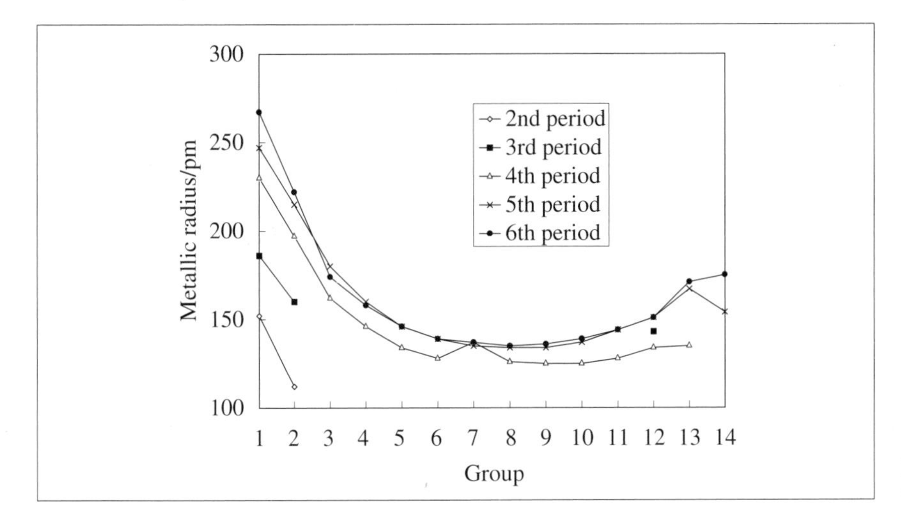

Figure 4.9 Metallic radii of the elements

The general decreases in metallic radii along the transition series may be called **transition metal contractions** and occur because of the poor shielding of the nuclear charge by the more diffuse d electrons (compared to those in s and p orbitals) offered to the additional d electrons; the effects are similar in the 4d and 5d series. The 3d orbitals are more compact, relative to the 4d and 5d sets, as pointed out in Section 3.1.3. The effect has consequences for the sizes of the subsequent elements in Groups 13–18, but they are more pronounced for those of Groups 13 and 14. The 3d contraction has an effect on the radius of gallium, as can be seen from Figure 4.10. The Na/Al variation is due to the general increasing effectiveness of the nuclear charge as the 3s and overlapping 3p bands are occupied by the 1–3 valence electrons. For the subsequent periods, the changes in radius from Group 2 to Group 13 include the 3d, 4d and 5d contractions in the respective periods and show a larger effect in period 4 such that the radius of gallium is slightly smaller than that of aluminium. This has an effect upon the chemical properties of gallium.

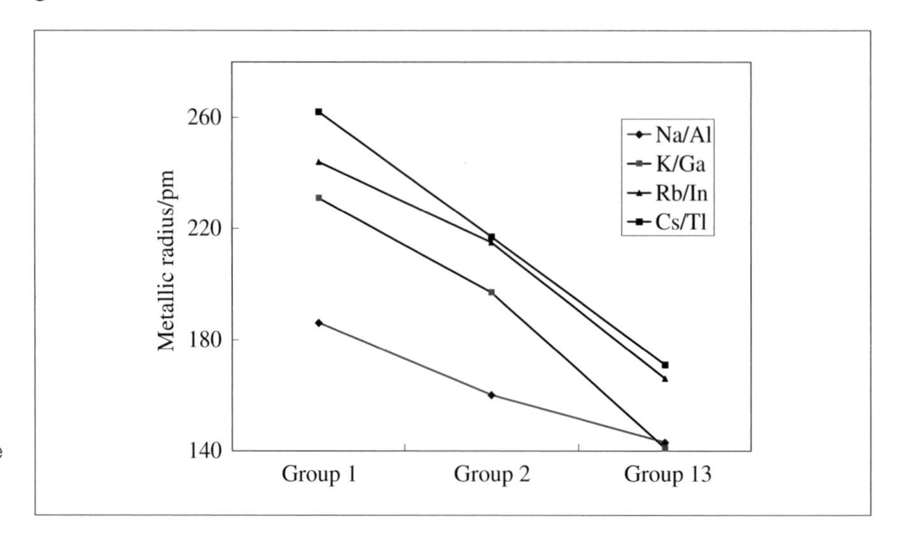

Figure 4.10 Metallic radii of the Groups 1, 2 and 13 elements of the 3rd, 4th, 5th and 6th periods

Size Variations of the Lanthanide and Actinide Elements

The lanthanide and actinide elements are all metallic and are formally members of Group 3 of the Periodic Table. They all form 3+ ions in their compounds and in aqueous solution, with few exceptions. The actinide elements form a larger range of oxidation states than the lanthanides, but the 3+ ions are used in this section for comparison purposes. Figure 4.11 shows the lanthanides' metallic and 3+ ionic radii. There is an almost regular decrease in metallic radius along the lanthanide series, with discontinuities at Eu and Yb. Most of the lanthanide metals contribute three electrons to bands of molecular orbitals con-

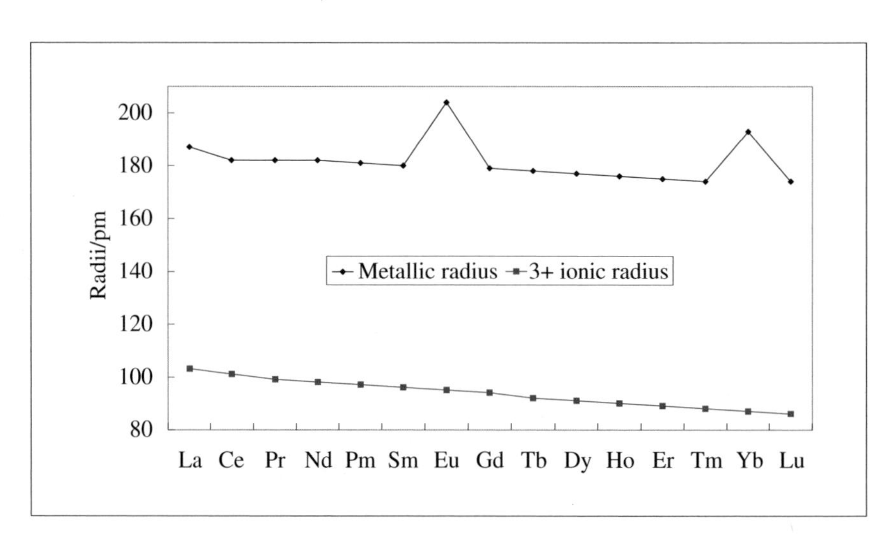

Figure 4.11 Metallic and 3+ ionic radii for the lanthanide elements

structed from the 5d and 6s atomic orbitals, their other valency shell electrons remaining as $4f^n$ configurations that do not interact with those on neighbouring atoms. In the cases of Eu and Yb, the atoms only contribute two electrons to the 5d/6s bands and retain the configurations $4f^7$ and $4f^{14}$, respectively, and so maximize their exchange energies of stabilization. The contributions of only two electrons each to the bonding bands makes the Eu and Yb metallic radii larger than those of their neighbours that contribute three electrons.

The 3+ ions have regular 4f orbital fillings, and there is an almost linear decrease in their ionic radii along the series. This is known as the **lanthanide contraction**, and has consequences for the transition elements of the sixth period. There are two contributing factors to the lanthanide contraction. One factor is that additional 4f electrons are not very efficient at shielding the nuclear charge, which becomes more effective along the series. The other factor is a relativistic effect which is described below. The lanthanide contraction is a major cause of the atoms of the transition elements of the sixth period being smaller than expected when compared to their group members in the fifth period, so that they are exceptions to the general rule that atoms become larger down any group because they have an extra shell of electrons to offer nuclear shielding.

The considerable reduction in radius when a metal loses three electrons is obvious from Figure 4.11 and occurs when any positive ion is formed. The remaining electrons are bound more tightly by the existing nuclear charge, which increases its effectiveness.

Figure 4.12 shows the limited amount of data for the metallic radii and 3+ ionic radii available for the actinide elements. The variation in metallic radii is due to varying numbers of electrons contributing to the 7s/6d bands, and to the varying structures of the actinides, which are complex. There is the expected reduction in size when three electrons are

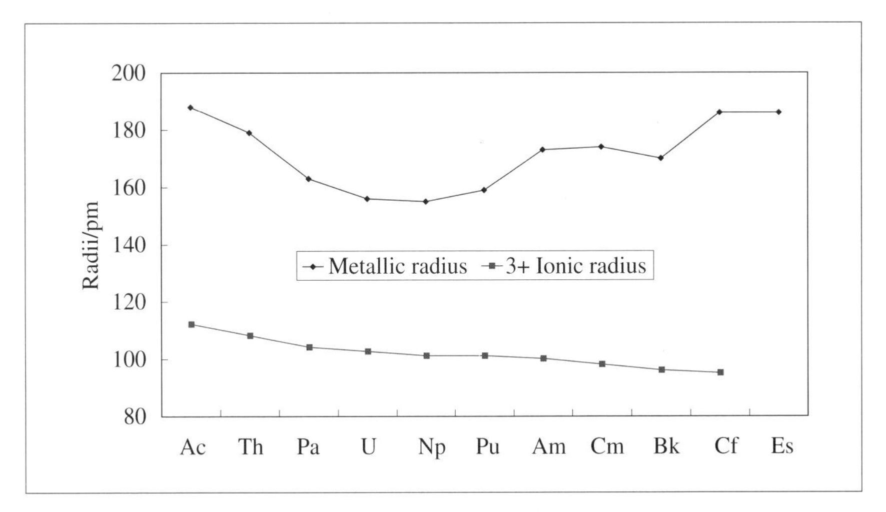

Figure 4.12 Metallic and 3+ ionic radii for the actinide elements

removed to give the 3+ ions and their radii do show an **actinide contraction** similar to that described for the lanthanides.

4.3.3 Variations in Covalent Radii

Figure 4.13 shows how covalent radii vary across periods and down the groups of the Periodic Table. Across periods there is a general reduction in atomic size, while down any group the atoms become larger. These trends are consistent with the understanding gained from the study of the variations of the first ionization energies of the elements. As the ionization energy is a measure of the effectiveness of the nuclear charge in attracting electrons, it might be expected that an increase in nuclear effectiveness would lead to a reduction in atomic size. The trends in atomic size in the Periodic Table are, in general, almost the exact opposite to those in the first ionization energy. The intermediate changes in the series of transition elements reflect those expected when the d electrons have to pair up and are no longer available for covalent bonding. Again, as is the case with metallic radii, the atoms of the fifth and sixth period transition elements are almost identical in size within each group. The size changes in Groups 1 to 13 reflect those of the Meyer curve of Figure 4.8, and thereafter there is a reduction in size across each period for Groups 14–17, opposite to the trends shown in the Meyer curve. This is because there are no relatively large contributions from van der Waals radii in the covalent radius plot, but they do contribute to the Meyer plot.

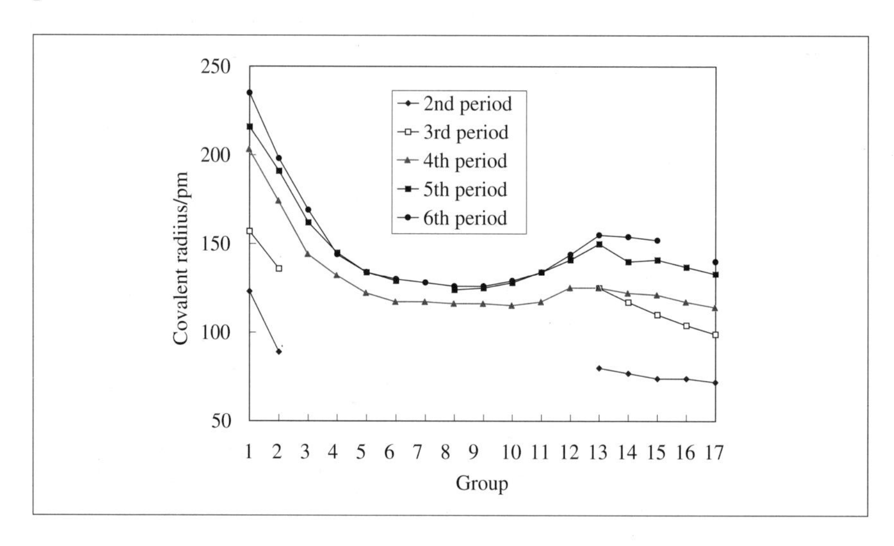

Figure 4.13 Covalent radii of the elements

From Figures 4.9 and 4.13 it is clear that an element's covalent radius is smaller than its metallic radius. This is because the element participating in covalent bonding is engaging in electron-pair bonding, which is stronger than the delocalized electron-band interactions occurring in the metallic state. These observations follow the generalization that there is an *inverse relationship* between the length and strength of chemical bonds.

Note the relatively large reduction in covalent radius between the first and second members of each main group. This is consistent with the greater binding experienced by the 2p electrons of the elements of the second period compared to the binding of the outer p orbitals of elements in subsequent periods.

4.3.4 Variations in Ionic Radii

Although ionic radii are only applicable to compounds rather than elements, a study of their trends in the Periodic Table provides insight into changes in the effectiveness of the nuclear charge as a function of the number of electrons removed to produce positive ions and the number added to give negative ions. Some selection is required in the data presented, to exemplify the trends across the periods when the charge on the ions varies and when the charge is constant down the groups. Table 4.4 gives the limited range of ionic radii for the ions produced by the main group metals and includes the ionic radii as percentages of the metallic radii of the elements.

Table 4.4 Some ionic radii which are also expressed as a percentage of the metallic radii of the element

Group 1	*Group 2*	*Group 13*	*Group 14*
Li^+, 60 (39%)	Be^{2+}, 31 (28%)		
Na^+, 95 (51%)	Mg^{2+}, 65 (41%)	Al^{3+}, 50 (35%)	
K^+, 133 (58%)	Ca^{2+}, 99 (50%)	Ga^{3+}, 62 (44%)	
Rb^+, 148 (61%)	Sr^{2+}, 113 (53%)	In^{3+}, 81 (49%)	Sn^{4+}, 71 (44%)
Cs^+, 169 (65%)	Ba^{2+}, 135 (62%)	Tl^{3+}, 95 (56%)	Pb^{4+}, 84 (48%)

The data in Table 4.4 show that there is a considerable decrease in the ionic radius across any period and an increase down any group. The changes are similar to those in the metallic and covalent radii, but are larger because ionic radii become smaller as the charge increases. The data are also expressed as a percentage of the metallic radius of each element. The percentages show that positive ion formation has a greater effect on the smaller elements and increases with increasing ionic charge. These observations are consistent with expected changes in the effectiveness of the nuclear charge. Across the periods, the ionic charge increases and the nuclear charge becomes more effective as there are fewer electrons to attract and there is less shielding of the outer electrons. Down the groups there is an increase in size, as expected from the greater shielding afforded by the extra shells of electrons.

There are only a few negative ions that participate in predominantly ionic compounds. Prewitt and Shannon have assigned ionic radii to the anions N^{3-}, S^{2-}, Se^{2-} and Te^{2-}, but compounds containing these elements in their particular oxidation states contain a significant fraction of covalent bonding. That only leaves three anions (O^{2-}, F^- and Cl^- with their respective ionic radii of 140, 136 and 181 pm) that take part in many compounds which are highly ionic in character. The ions are significantly larger than their neutral atoms. The respective covalent radii are 74, 72 and 99 pm. That anions are larger than neutral atoms is expected from considerations of reduction of the effectiveness of the nuclear charge and increased interelectronic repulsion.

The variations in ionic radii of the transition elements of the fourth period serve to exemplify the arguments needed to rationalize similar variations in the other transition series. Figure 4.14 shows a plot of the radii of the 2+ ions of those transition elements of the fourth period which form them.

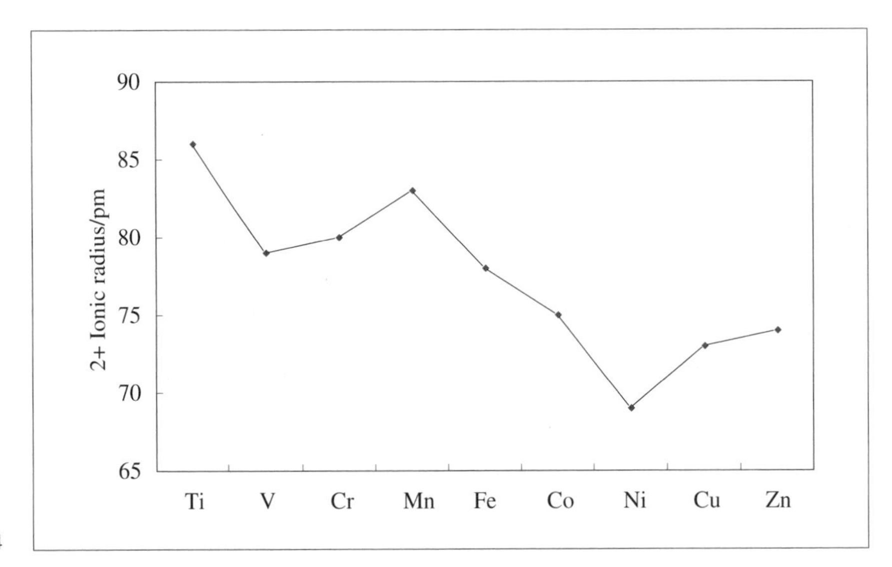

Figure 4.14 Ionic radii of the 2+ transition elements of Period 4

The radii are those of octahedrally coordinated ions as they are found in crystalline compounds, the counter-ions (*i.e.* the ions of opposite charge) being situated at the vertices of an octahedron, as shown in Figure 4.15.

There is a general downward trend in the radii going across the period, but the dips at V and Ni and the shape of the plot may be explained in terms of which d orbitals are occupied in each case. Regarding the sequence in terms of the progressive filling of the 3d orbitals, it is assumed that the first d orbitals to be filled are the $3d_{xy}$, $3d_{xz}$ and $3d_{yz}$ set, followed by the $3d_{z^2}$ and $3d_{x^2-y^2}$ set, and that the counter-ions are placed along the coordinate axes as shown in Figure 4.15.

The spatial arrangements of the 3d orbitals are shown in Figure 2.8.

The electrons in the $3d_{xy}$, $3d_{xz}$ and $3d_{yz}$ set are placed so that they bisect the lines between the nucleus of the cation and the counter-ions

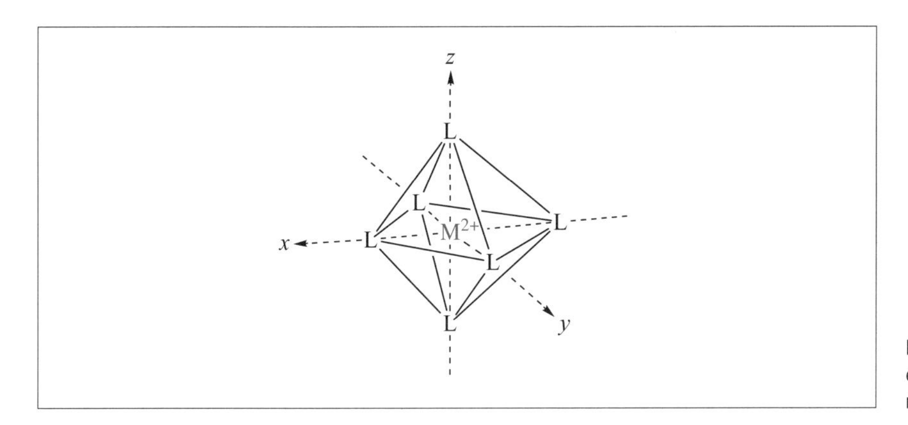

Figure 4.15 The octahedral environment of a 2+ transition metal ion

and offer minimum shielding of the nucleus. Thus, adding two electrons to the $3d_{xy}$, $3d_{xz}$ and $3d_{yz}$ set of orbitals allows the Ti nucleus to exert maximum attractive force on the counter-ions, and the ionic radius of Ti^{2+} is smaller than it might otherwise have been. The addition of an extra electron in V^{2+} causes the $3d_{xy}$, $3d_{xz}$ and $3d_{yz}$ set of orbitals to be singly occupied in accordance with Hund's rules, and the inefficient shielding of the nucleus produces a decrease in the ionic radius as compared to that of Ti^{2+}. The next orbital to be filled in Cr^{2+} is one of the $3d_{z^2}$ and $3d_{x^2-y^2}$ set. These orbitals are localized along the coordinate axes, and electrons in them offer maximum shielding to the counter-ions. Any electrons occupying the $3d_{z^2}$ and $3d_{x^2-y^2}$ set of orbitals tend to allow an increase in the ionic radius of the cation. This occurs with Cr^{2+} and Mn^{2+}, but when all five 3d orbitals are singly occupied the sixth electron must pair up; the pattern is repeated in the next five 2+ ions with the $3d_{xy}$, $3d_{xz}$ and $3d_{yz}$ set of orbitals being filled progressively in Fe^{2+}, Co^{2+} and Ni^{2+}, leading to a decrease in ionic radius. The last two ions, Cu^{2+} and Zn^{2+}, show an increase in ionic radius because the added electrons occupy the highly shielding $3d_{z^2}$ and $3d_{x^2-y^2}$ set of orbitals. The overall reason why the 3d orbitals are filled in the manner described is because such filling produces greater stability of the compounds formed than would alternative fillings.

4.4 Electronegativity

The concept of electronegativity is derived from experimental observations, such that the elements fluorine and chlorine are highly electronegative in their strong tendencies to become negative ions. The metallic elements of Group 1, on the other hand, are not electronegative and are better described as being electropositive: they have a strong tendency to form positive ions. A scale of electronegativity coefficients would be useful in allowing a number to represent the tendency of an element in a molecule to attract electrons to itself. The establishment of such a scale has involved the powers of two Nobel Chemistry Prize winners (Pauling in 1954, for valence bond theory, and Mulliken in 1966, for molecular orbital theory), but, after many other efforts, there are still doubts about the currently accepted values and about their usefulness.

4.4.1 The Pauling Scale of Electronegativity Coefficients

Pauling suggested that the bond in a heteronuclear molecule, AB, would be stronger than that expected for a purely covalent bond owing to what he termed as ionic–covalent resonance energy. This is to be interpreted by considering the bonding in AB as some mixture of a covalent bond, A–B, and an ionic form, A^+B^- (assuming that B is more electronegative than A). This is the language of valence bond theory.

Pauling's scale of electronegativity was based on the observation that the bond dissociation energy of a diatomic molecule, AB, was normally greater than the average values for the diatomic molecules, A_2 and B_2, for any pair of elements, A and B. It is important to note that the three molecules AB, A_2 and B_2 were those in which only **single covalent bonds** existed.

He stipulated that the geometric mean of the bond dissociation energies of the molecules A_2 and B_2 represented the strength of a *purely* covalent bond in the molecule AB, and that any extra strength of the A–B bond was because of the of the difference in electronegativity of the two atoms. This extra strength is represented by the equation:

$$\Delta = D(\text{A–B}) - [D(\text{A}_2) \times D(\text{B}_2)]^{1/2} \quad (4.4)$$

Pauling found that there was a correlation between the square root of Δ and the positive differences between values of the electronegativity coefficients of the two elements concerned:

$$\Delta^{1/2} \propto |\chi_\text{A} - \chi_\text{B}| \quad (4.5)$$

By assigning the value of 2.1 for the electronegativity coefficient of hydrogen (this was done to ensure that all values of the coefficients would be positive), Pauling was able to estimate values for a small number of elements for which the thermochemical data were available. The scale was limited by lack of data and by the perversity of the majority of elements in not forming diatomic molecules.

4.4.2 Mulliken's Scale of Electronegativity Coefficients

Mulliken's scale was based on the logic that electronegativity is a property of an element that can be related to the magnitude of its first ionization energy, I_1, and to that of its first electron attachment energy, E_1, both being indications of the effectiveness of the nucleus in attracting electrons. Ionization energy is a measure of the effectiveness of a nucleus in retaining electrons and electron attachment energy is a measure of the attraction of the neutral atom for an incoming electron. The Mulliken coefficients were calculated as the average of the first ionization energy and the energy *released* when the atom accepted an electron (*i.e.* the value of the first electron attachment energy with its sign changed): $\frac{1}{2}(I_1 - E_1)$, with a suitable modifying constant to ensure the best correlation with the already established Pauling values.

An electronegative atom is expected to have a high first ionization energy and to have a large negative value for its first electron attachment energy, and to arrange for the two energies to work in the same direction it is necessary to change the sign of the latter.

The scale was limited by the available accurate data, particularly the values of the electron attachments energies. These latter are now well established, but the up-dated Mulliken scale has not found general acceptance. The correlation with the Allred–Rochow scale (Section 4.4.3, below) is statistically highly significant, with a correlation coefficient of 0.9. The Mulliken values are, on average, 8% lower than the corresponding Allred–Rochow values.

4.4.3 The Allred–Rochow Scale of Electronegativity Coefficients

Of several attempts at constructing an accurate scale of electronegativity, the one derived by Allred and Rochow[4] is now generally accepted and is known by their names. A discussion of its accuracy is included below. The scale is based on the concept that the electronegativity of an element is related to the force of attraction experienced by an electron at a distance from the nucleus equal to the covalent radius of the particular atom. According to Coulomb's law this force is given by:

$$F = \frac{Z_{\text{eff}} e^2}{r_{\text{cov}}^2} \tag{4.6}$$

where Z_{eff} is the effective atomic number and e is the electronic charge. The effective atomic number is considered to be the difference between the actual atomic number, Z, and a shielding factor, S, which is estimated by the use of **Slater's rules**. These represent an approximate method of calculating the screening constant, S, such that the value of the effective atomic number, Z_{eff}, is given by $Z - S$. The value of S is obtained from the following rules:

1. The atomic orbitals are divided into groups: (1s); (2s, 2p); (3s, 3p); (3d); (4s, 4p); (4d); (4f); (5s, 5p); (5d); (5f); and so on.
2. For an electron (principal quantum number n) in a group of s and/or p electrons, the value of S is given by the sum of the following contributions:
   (i) Zero from electrons in groups further away from the nucleus than the one considered.

(ii) 0.35 from each other electron in the same group, unless the group considered is the 1s when an amount 0.30 is used.
(iii) 0.85 from each electron with principal quantum number of $n - 1$.
(iv) 1.00 from all other inner electrons.

3. For an electron in a d or f group the parts (i) and (ii) apply, as in rule 2, but parts (iii) and (iv) are replaced with the rule that all the inner electrons contribute 1.00 to S.

Slater's rules are used to calculate the values of the Allred–Rochow electronegativity coefficients. In the calculations of the screening constants for the elements, Allred and Rochow applied Slater's rules to *all* the electrons of the atoms. In normal use of the rules, one of the outermost electrons is chosen to be the one "feeling the force". Applying Slater's rules in the *normal* way leads to a best fit with the older accepted Pauling/Mulliken values for electronegativity coefficients, given by the equation:

$$\chi = \frac{3590(Z_{\text{eff}} - 0.35)}{r_{\text{cov}}^2} + 0.744 \quad (4.7)$$

in which r_{cov} is expressed in picometres.

Worked Problem 4.6

Q The covalent radii of the F, N and Cl atoms are 72, 74 and 99 pm, respectively. Calculate the Allred–Rochow electronegativity coefficients for these elements.

A The electronic configurations of the atoms are: F, $1s^22s^22p^5$; Cl, $1s^22s^22p^63s^23p^5$; N, $1s^22s^22p^5$. The screening constant for F is calculated as 6 × 0.35 for six of the $2s^22p^5$ electrons, plus 2 × 0.85 for the two inner electrons, making a total of 3.8. The atomic number of F is 9, so the effective nuclear charge is 9 – 3.8 = 5.2. The screening constant for N is calculated as 4 × 0.35 for four of the $2s^22p^3$ electrons, plus 2 × 0.85 for the two inner electrons, making a total of 3.1. The atomic number of N is 7, so the effective nuclear charge is 7 – 3.1 = 3.9. The screening constant for Cl is calculated as 6 × 0.35 for six of the $3s^23p^5$ electrons, plus 8 × 0.85 for the eight $2s^22p^6$ next innermost electrons, plus 2 for the $1s^2$ pair, making a total of 10.9. The atomic number of Cl is 17, so the effective nuclear charge is 17 – 10.9 = 6.1. Placing these values for the effective nuclear charge into equation (4.7) gives the electronegativity coefficients of 4.10 for F, 3.07 for N and 2.85 for Cl.

The values given in modern versions of the Periodic Table and in data books are the Allred–Rochow values rounded off (usually) to one decimal place. To attempt to represent χ more accurately would be imprudent after considering the difficulties in obtaining any values at all. The method of calculation offers the possibilities of assigning χ values to the different oxidation states of the same element. It would be expected, for instance, that the electronegativities of manganese in its oxidation states of 0, +2, +3, +4, +5, +6 and +7 would be different, and providing that the necessary covalent radii are known the χ values are easily calculated.

Figure 4.16 shows the variation of Allred–Rochow electronegativity coefficients for singly bonded elements along periods and down the s- and p-block groups of the Periodic Table. In general, the value of the electronegativity coefficient increases across the periods and decreases down the groups. That is precisely the opposite of the trends in covalent radii, but similar to the trends in first ionization energy (Figure 4.1). This latter conclusion is no surprise, as it is to be expected that elements with a high tendency to attract electrons possess high first ionization energies. The period 4 elements of Groups 13–16 (Ga, Ge, As and Se) show significant exceptions to the general deceasing trend in electronegativity coefficient values down the groups. This is because of the 3d contraction, which causes alternation of some properties of the Groups 13–16 elements which are referred to in Chapter 5.

The electronegativity coefficients of the transition elements vary over the small range 1.1–1.7 and are not particularly useful in rationalizing

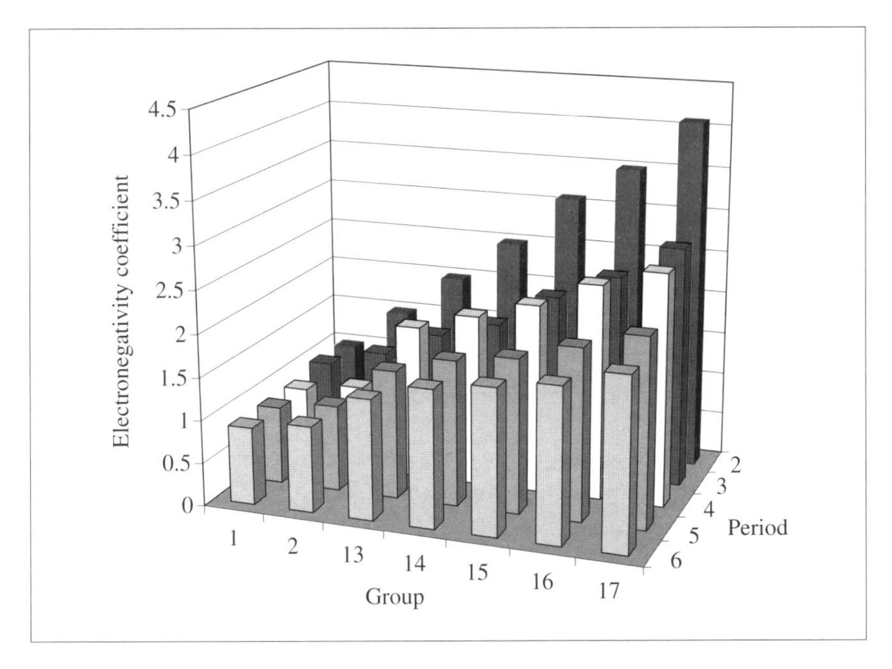

Figure 4.16 The Allred–Rochow electronegativity coefficients of the main group elements of periods 2–6

the properties of those elements, given the inherent difficulties and uncertainties in the derivation of their values. The lanthanide elements all have the same value of 1.1 for their electronegativity coefficients, which expresses fairly accurately their very electropositive nature, coming between magnesium (1.2) and calcium (1.0).

There have been attempts to relate differences in electronegativity coefficients to the percentage ionic character of bonds between different elements. These are reasonable on a qualitative level, but quantitative equations are found only to apply roughly to particular series of compounds (*e.g.* H–F to H–I), no broader generalizations being reasonable.

Possible Errors in Estimating Electronegativity Coefficients

Reference to the deficiencies of the Pauling and Mulliken scales of electronegativity are mentioned above. Possible errors in the estimation of the Allred–Rochow values are those associated with equation (4.7): the estimation of the effective nuclear charge and the determination of the covalent radius of the particular element. As indicated above, covalent radii are chosen to give a self-consistent set of values, and in any particular case could be in error. In using this set of covalent radii there may be errors for each element, but they are likely to be consistent errors which do not alter the electronegativity coefficient values by so much as to alter the order of elements when placed in an electronegativity scale. The main errors arise by the use of the very approximate Slater's rules for the determination of the shielding constants for the elements and thence their effective nuclear charges.

There have been attempts to obtain better values for effective nuclear charges, in particular those of Clementi and Raimondi and of Froese and Fischer. As yet, these have not been used for the construction of electronegativity scales for all the elements.

One of the most debatable inconsistencies in the order of electronegativity coefficients of the elements is the odd position of nitrogen (3.1) with respect to chlorine (2.8). That nitrogen is more electronegative than chlorine is not believed by chemists who know their chemistry! For example, chlorine reacts with the transition elements to give high oxidation states, *e.g.* tungsten and rhenium forms +6 chlorides, whereas the transition elements react at high temperatures with dinitrogen to give compounds of the formula MN, where M represents the transition element. The nitrogen compounds conduct electricity and are best described as interstitial compounds, *i.e.* the nitrogen atoms occupy some of the interstices or interatomic spaces in the metal lattice and do not exist as nitride ions, N^{3-}. Ionic nitrides do exist, but are formed only by the very electropositive elements, Li, Mg, Ca, Sr, and Ba and Zn. In the compound NCl_3 it is the nitrogen atom which is oxidized to its +3 state by the *more electronegative* chlorine.

4.4.4 Electronegativity Coefficients and Types of Bonding

A generally useful application of electronegativity coefficients is embodied in what is known as the van Arkel–Ketelaar triangle. This is shown in outline in Figure 4.17 for the elements caesium and fluorine and their ionic reaction product CsF, and is a plot of the difference between the electronegativity coefficients of a binary compound against their mean value. It summarizes well the differences in bonding at the three extremes of the triangle, as indicated in Table 4.5.

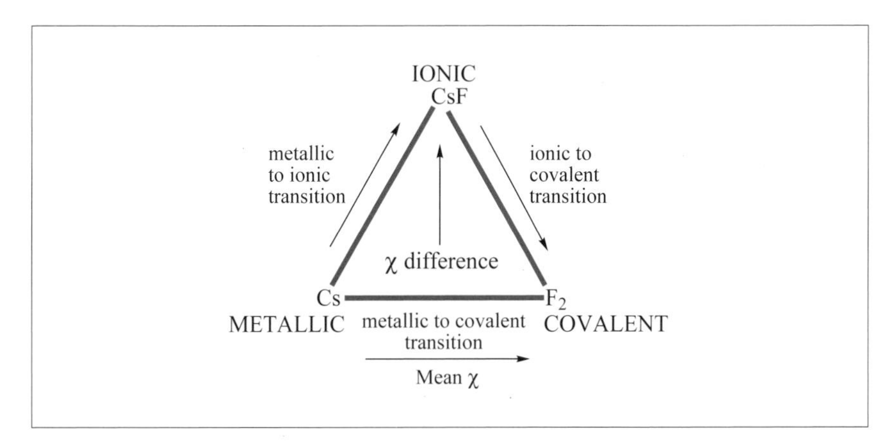

Figure 4.17 A Van Arkel–Ketelaar triangle

Table 4.5 Types of bonding related to electronegativity coefficients of the participating elements

Mean value of χ	$\Delta\chi$	*Type of bonding*
Small	Zero	Metallic
Intermediate	Large	Ionic
Large	Zero	Covalent

Along the base of the triangle, for which $\Delta\chi$ is zero, are the elements in which the bonding varies from metallic to covalent as the value of χ increases. Along the left-hand side of the triangle is the transition from metallic bonding to ionic as $\Delta\chi$ increases, and along the right-hand side there is the transition from ionic bonding to covalent as $\Delta\chi$ decreases.

4.5 Relativistic Effects

The group and periodic trends of ionization energies, atomic sizes and electronegativity coefficients are discussed above in terms of the variations in electronic configurations of the atoms. The values of these properties are influenced by **relativistic effects**[5,6] which, for valence electrons, increase with the value of Z^2, and become sufficiently important in the elements of the sixth period (Cs to Rn) to explain largely their chemical differences from the elements of the fifth period (Rb to Xe).

The theory of relativity expresses the relationship between the mass m of a particle travelling at a velocity v and its rest mass, m_0:

$$m = \frac{m_0}{\left(1 - \frac{v^2}{c^2}\right)^{1/2}} \quad (4.8)$$

where c is the velocity of light. The average velocity of a 1s electron, in *atomic units*, is given by the atomic number (Z) of the element. That of a 1s electron of an atom of mercury (Z = 80) is thus 80/137 of the velocity of light (see Box 4.3 on atomic units).

Equation (4.8) contains the possible implication that element 137 would be the last in the series of elements, since any higher values of Z would have 1s electron velocities greater than that of light. In a personal communication to the author, Professor Pekka Pyykö wrote: "The element 138 would be in trouble, IF the nuclei were points. In the exact Darwin–Gordon (1928) solution of the Dirac equation for a Coulomb potential, the lowest, 1s, eigenvalue (*i.e.* orbital energy) would dive into the positron-like continuum if $Z > c$. A realistic finite nuclear size prevents this up to about Z = 160 or so."

Box 4.3 Atomic Units

Atomic units are used in relativistic calculations. Velocities are normally expressed as metres per second, but the atomic units of length and time are very different. The atomic unit of length is the Bohr radius of the lowest energy hydrogen orbit, a_0, which has the value 52.9177249 pm. The atomic unit of time is the time it takes an electron in the lowest energy orbit of the hydrogen atom to travel a distance equal to one Bohr radius, $2\pi m a_0^2/h$, which is 2.41888×10^{-17} s. The normal value for the velocity of light, c, is 299792458 m s^{-1} and this is expressed in atomic units as: $c(\text{m s}^{-1}) \div [a_0(\text{m}) \div 2.41888 \times 10^{-17}(\text{s})] = 137.036$. The velocity of an electron in the 1s orbital of an atom is given in atomic units by the value of Z, the atomic number.

Equation (4.8) gives an estimate of the mass of the mercury 1s electron as about 23% greater than its rest mass. Since the radius of the 1s orbital is inversely proportional to the mass of the electron, the radius of the orbital is reduced by about 23% compared to that of the non-relativistic radius. This s orbital contraction affects the radii of all the other orbitals in the atom up to, and including, the outermost orbitals. The s orbitals contract, the p orbitals also contract, but the more diffuse d and f orbitals can become more diffuse as electrons in the contracted s and p orbitals offer a greater degree of shielding to any electrons in the d and f orbitals.

The effects are observable by a comparison of the metallic radii and the first two ionization energies of the elements of Group 12 (Zn, Cd and Hg), as shown in Table 4.6. The mercury atom is smaller than expected from an extrapolation of the zinc–cadmium trend and is more difficult to ionize than the lighter atoms. In consequence, the metal–metal bonding in mercury is relatively poor, resulting in the element being a liquid in its standard state at 298 K. This almost Group 18 behaviour of mercury may be compared to that of the real Group 18 element, xenon, which has first and second ionization energies of 1170 and 2050 kJ mol^{-1}.

Table 4.6 The metallic radii and first two ionization energies of the Group 12 elements

Atom	*r*/pm	I_1/kJ mol^{-1}	I_2/kJ mol^{-1}
Zn	133	908	1730
Cd	149	866	1630
Hg	152	1010	1810

Relativistic effects are also observable in the properties of the Group 11 elements (Cu, Ag and Au). Their metallic radii, first ionization energies and their first electron attachment energies are given in Table 4.7. The closeness of the metallic radii of Ag and Au is a relativistic effect. There is now clear experimental evidence that Au^{I} is smaller than Ag^{I}: in comparable covalent compounds, their covalent radii are 137 and 146 pm, respectively.[7] The relatively high value of the first ionization energy of gold is a major factor in the explanation of the nobility of the metal. The extraordinarily highly negative electron attachment energy of gold (compare this value with that for fluorine, –322 kJ mol^{-1}) is consistent with the existence of the red crystalline ionic compound $Cs^{+}Au^{-}$, which is produced by heating together an equimolar mixture of the two elements, gold behaving like a halogen!

Platinum also has, for a metal, a very large negative value of electron attachment energy of –247 kJ mol^{-1}.

The halogen-like behaviour of gold was first noted by R. S. Nyholm.

Table 4.7 The metallic radii, first ionization energies and first electron attachment energies of the Group 11 elements (energies in kJ mol^{-1})

Atom	*r*/pm	I_1	*E*
Cu	128	745	–119
Ag	144	732	–126
Au	144	891	–223

The general group similarities exhibited by members of the second and third transition series are due to their almost identical atomic radii. The reason for the anomalously low sizes of atoms of the third transition series is usually given in terms of the lanthanide contraction. As the 4f set of orbitals is filled in the elements from lanthanum to lutetium there is a contraction in atomic radius from 188 pm (La) to 174 pm (Lu). This contraction is regarded as having no more than a 14% relativistic contribution, but is responsible for ~50% of the apparent contraction of the transition metals of the third series. The other ~50% is because of the increasing relativistic contraction of the transition elements themselves, due to their high Z values, the maximum effect occurring in gold.

The exceptions to the general trends of first ionization energies across periods and down groups, mentioned in Sections 4.1.1 and 4.1.4, are explicable when relativistic effects are considered. Thallium and lead have higher values of their first ionization energies than expected from the trends down their respective groups because their p orbitals are more compact. The relativistic effect upon the 6p orbitals of the elements from Tl to Rn is to reinforce a stabilization of one orbital with respect to the other two (see Box 4.4 on Spin–Orbit Coupling). Instead of the expected trend, the first ionization energies of Tl, Pb and Bi (589, 715 and 703 kJ mol^{-1}) do not show a general increase like those of In, Sn and Sb, (558, 709 and 834 kJ mol^{-1}): the value for Bi is lower than that of Pb.

Another notable difference in properties down groups is the "**inert pair effect**", as demonstrated by the chemical behaviour of Tl, Pb and Bi. The main oxidation states of these elements are +1, +2 and +3, respectively, which are lower by two units than those expected from the behaviour of the lighter members of each group. There is a smaller, but similar, effect in the chemistry of In, Sn and Sb. These effects are partially explained by the relativistic effects on the appropriate ionization energies, which make the achievement of the higher oxidation states (the participation of the pair of s electrons in chemical bonding) relatively more difficult.

Box 4.4 Spin–Orbit Coupling

The interactions of the orbital and spin momenta (spin–orbit coupling) contribute to the inert pair effect by increasing the difficulty of promoting s electrons to p orbitals. Although the coupling of the angular and spin momenta is beyond the scope of this text, a brief mention of the subject is appropriate. The electronic states of the lighter elements are best described by the **Russell–Saunders (RS) coupling** scheme in which the individual l values of the valence electrons combine to give values of the **total angular momentum quantum number**, L, and the individual spin quantum numbers combine to give a **total spin momentum quantum number**, S. The RS scheme, which governs the interactions between the L and S

quantum numbers, is inadequate for the heavier elements. The electronic states of the heavier elements are better described by what is known as **j–j coupling**. In this scheme, the quantum numbers l and s for individual electrons combine to give a "small j" value for each electron. The rule for this combination is that:

$$j = l + s,\ l + s - 1,\ \ldots\ |l - s| \tag{4.9}$$

With regard to an external magnetic field, the corresponding m_j values are given by:

$$m_j = j,\ j - 1,\ \ldots\ -(j - 1),\ -j \tag{4.10}$$

The j–j coupling leads to a breakdown of the normal triple degeneracy of the 6p orbitals. This produces one orbital at lower energy that can contain electrons with $j = \frac{1}{2}$ and with m_j values of $\frac{1}{2}$ and $-\frac{1}{2}$, and is labelled by its j value of $\frac{1}{2}$ as $p_{1/2}$. The other two orbitals, at higher energy than the $p_{1/2}$, are doubly degenerate, and have $j = \frac{3}{2}$ and m_j values of $\frac{3}{2}$, $\frac{1}{2}$, $-\frac{1}{2}$ and $-\frac{3}{2}$, and are labelled as $p_{3/2}$.

The two electrons in the low-energy $s_{1/2}$ orbital (with $j = \frac{1}{2}$ and with m_j values of $+\frac{1}{2}$ and $-\frac{1}{2}$) are **relativistically stabilized** with respect to the p levels, and form the "**inert pair**" typical of the chemistry of the 6p elements. In thallium the single 6p electron is in the $p_{1/2}$ orbital and to achieve trivalency a promotion of one of the 6s electrons to the relatively higher energy $p_{3/2}$ orbital is necessary. In the lead atom, the two 6s electrons occupy the $s_{1/2}$ orbital. The $p_{1/2}$ orbital is doubly occupied in lead and in bismuth. The single occupation of the destabilized $p_{3/2}$ orbital in bismuth explains the observation that the first ionization energy of the element (703 kJ mol^{-1}) is lower than that of lead (715 kJ mol^{-1}).

Summary of Key Points

1. The periodic manner (periodicity) in which the first ionization energies of the elements, successive ionization energies of some selected elements, and the electron attachment energies of the first 36 elements vary as the atomic number increases was described.

2. Atomic size was defined in terms of covalent, metallic and van

der Waals radii, and variations in atomic sizes across periods and down groups were discussed.

3. Definitions of electronegativity coefficients were described, and variations of electronegativity coefficients across periods and down groups were rationalized.

4. Relativistic effects on orbital energies, and atomic sizes and properties were described.

Problems

4.1. The data below consist of successive ionization energies (in kJ mol^{-1}) for three elements. Identify the group of the Periodic Table to which each element belongs, and explain the variations in the values for each element.

Element	I_1	I_2	I_3	I_4	I_5
A	1090	2350	4610	6220	37800
B	577	1820	2740	11600	14800
C	376	2420	3300	4400	6000

4.2. The ionic radii of P^{V} and Mn^{VII} are quoted as 58 pm and 39 pm, respectively, for tetrahedral coordination. The P–O bond length in the tetrahedral phosphate ion, PO_4^{3-}, is 154 pm, the Mn–O bond length in the tetrahedral manganate(VII) ion (permanganate ion), MnO_4^-, is 163 pm. The ionic radius of the oxide ion is 140 pm. Why are the actual bond lengths in the two ions less than the sums of the appropriate ionic radii? The covalent radii of P and Mn, for single bonds, are 110 pm and 117 pm, respectively, and that of oxygen is 74 pm. What do these figures imply for the state of bonding in the two ions?

4.3. The first and second ionization energies of the Group 1 elements are given as:

Group 1 element	*First ionization energy*/kJ mol^{-1}	*Second ionization energy*/kJ mol^{-1}
Li	519	7300
Na	494	4560
K	418	3070
Rb	402	2650
Cs	376	2420

Explain: (i) the decrease in the values of the first ionization energies down the group, (ii) the very high values of the second ionization energies, and (iii) the decrease in the values of the second ionization energies down the group, in terms of the electronic changes that occur.

4.4. Although trends in the densities of the elements is not a topic treated in this book, the variation of density down a group of the Periodic Table may be dealt with in simple terms by carrying out the following exercise. The RAM values, densities and metallic radii of the Group 2 elements are given as:

Element	*RAM*	*Density*/ kg m^{-3}	r_M/pm	*Crystal structure*
Be	9	1850	112	hcp
Mg	24	1740	160	hcp
Ca	40	1540	197	ccp
Sr	87.6	2620	215	ccp
Ba	137.3	3510	217	bcc

hcp = hexagonal closest packing; ccp = cubic closet packing; bcc = body-centred cubic. See Section 5.1.2.

The crystal structures hcp, ccp and bcc, if considered to consist of hard sphere atoms, are filled to the extent of 74%, 74% and 68%, respectively. (i) Plot the densities of the elements against their RAM values to emphasize the variation down the group. (ii) Explain the variation by calculating the densities of the elements from the data given. The density depends upon the intrinsic mass of the atoms of the solid, which is proportional to the RAM value and depends upon the crystal structure which indicates the efficiency of space-filling by the supposed hard-sphere atoms. The density is given by the equation:

$$\text{Density} = \frac{\text{RAM} \times 10^{-3} \times f \times 3}{4\pi \times N_A \times r_M^3}$$

In the equation, f represents the fraction of space filled by the hard-sphere atoms. Compare the calculated densities with the values quoted. The discrepancies are small, and are due to errors in the metallic radii and in the assumption that atoms are hard spheres.

References

1. D. I. Mendeleev, *J. Russ. Chem. Soc.*, 1869, iC, 60–77.
2. J. L. Meyer, *Annalen*, 1870, suppl. vii, 354.
3. R. D. Shannon, *Acta Crystallogr., Sect A*, 1976, **32**, 751.
4. A. L. Allred and E. G. Rochow, *J. Inorg. Nucl. Chem.*, 1958, **5**, 264.
5. N. Kaltsoyannis, *Relativistic effects in inorganic and organometallic chemistry*, in *J. Chem. Soc., Dalton Trans.*, 1997, 1. This review paper covers relativistic effects very well, and contains 88 references to the original literature on the basis of the effects and on specific cases.
6. P. Pyykkö, *Relativistic effects in structural chemistry*, in *Chem. Rev.*, 1988, **88**, 563. A review paper containing 428 references and written by the master of the subject.
7. U. M. Tripathi, A. Bauer and H. Schmidbaur, *J. Chem. Soc., Dalton Trans.*, 1997, 2865. On the covalent radii of Ag^{I} and Au^{I}.

Further Reading

R. J. Puddephatt and P. K. Monaghan, *The Periodic Table of the Elements*, 2nd edn., Oxford University Press, Oxford, 1986. A concise description of the structure of the Periodic Table and a discussion of periodic trends of many physical and chemical properties of the elements.

D. M. P. Mingos, *Essential Trends in Inorganic Chemistry*, Oxford University Press, Oxford, 1998. A new look at trends in physical and chemical properties of the elements and their compounds. This is a general text that would cover the material of most university courses in inorganic chemistry, apart from some specialized topics, but is recommended here for the sections on atomic structure.

5

Periodicity II: Valencies and Oxidation States

This chapter deals with the periodicity of the valencies and oxidation states of the elements. An overview of chemical bonding is included to aid in the understanding of the periodicities of the enthalpies of atomization of the elements, and the nature and properties of their fluorides and oxides, which are the subjects of Chapters 6 and 7.

Aims

By the end of this chapter you should understand:

- The basis of covalent, metallic and ionic bonding
- The difference between valency and oxidation state
- The variations of valencies and oxidation states of the elements across periods and down groups
- The octet rule and exceptions from it
- The 18-electron rule and exceptions from it

5.1 An Overview of Chemical Bonding

The details of chemical bonding are left to other books in this series, but it is useful in this text to summarize the main features of the ways in which atoms can bond to each other in elementary forms and in compounds. In the pure elements, the bonding may be described as either metallic or covalent, with small covalent molecules held together by van der Waals intermolecular forces in their liquid and solid states. In addition, because compounds are discussed at some length in subsequent chapters, the basis of ionic bonding is described.

5.1.1 Covalent Bonding

Covalency results from the formation of **electron-pair bonds**, in which both of the two participating atoms supply one electron; the electron pair is shared between the two atoms and constitutes a **single covalent bond**. The simplest electron-pair bond is that in the dihydrogen molecule, in which the two 1s valency electrons of the hydrogen atoms pair up in a **sigma (σ) bonding molecular orbital**. This interaction is shown in Figure 5.1

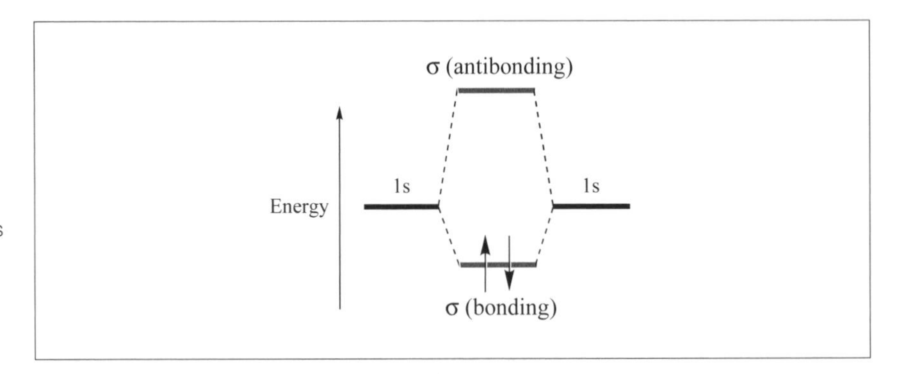

Figure 5.1 The relative energies of the anti-bonding and bonding molecular orbitals of the dihydrogen molecule and the hydrogen 1s orbital

As atomic orbitals are labelled s, p and d, molecular orbitals are labelled σ, π and δ, depending on the type of overlap between the atomic orbitals from which they are constructed.

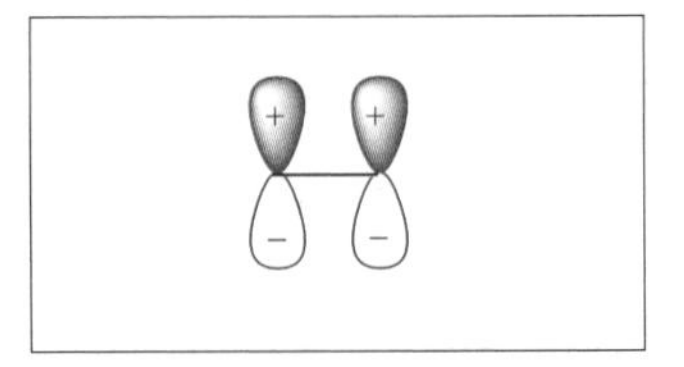

Figure 5.2 The π overlap of two 2p atomic orbitals

The bonding orbital is cylindrically symmetric along the molecular axis, and such orbitals are termed *sigma* to distinguish them from other symmetries which may occur. Depending on their valency, some atoms form either more than one single bond or participate in **multiple bonding**. **Pi (π) bonding** occurs when p orbitals overlap sideways as shown in Figure 5.2.

For example, the bonding in the dinitrogen molecule can be considered to consist of one σ bond and two π bonds, giving an overall bond order of three; *i.e.* it can be thought of as a triple bond.

Coordinate or **dative bonding** is a special type of covalency in which both electrons forming the electron-pair bond are supplied by one of the participating atoms. This kind of bonding is important in **adducts**, *e.g.* that between BF_3 and NH_3 in which the lone pair of electrons on the nitrogen atom is donated to a 2p orbital of the boron atom, $H_3N{:}\rightarrow BF_3$. The most widespread use of coordinate bonding is in **complexes** of the transition elements, in which the central metal has a particular oxidation state, but can accept six or even more electron pairs from suitable ligands which may be negative ions or neutral molecules. For example, the complex ion hexacyanoferrate(III), $[Fe(CN)_6]^{3-}$, can be visualized as a central Fe^{3+} ion (hence the III in its name) surrounded by six cyanide ions, each of which provides a pair of electrons to make the six coordinate bonds.

5.1.2 Metallic Bonding

This occurs in elements which exist in the metallic state, with each atom surrounded by more neighbours than it has electrons available to form electron-pair bonds. To maximize the bonding effect of the valency electrons, they occupy **bands** of molecular orbitals which extend throughout the metal and allow good electrical conduction to occur. Such bonding can only operate in the solid and liquid states when each atom has sufficient neighbours to allow the delocalization of the electrons into the bands.

Metal atoms can only form weak covalent bonds with each other, and preferentially exist in the *solid state* with **crystal lattices** in which the **coordination number** (the number of nearest neighbours) of each atom is relatively high (8–14). Collaboration between many atoms is required for metal stability. Diagrams of the three most common metallic lattices are shown in Figure 5.3.

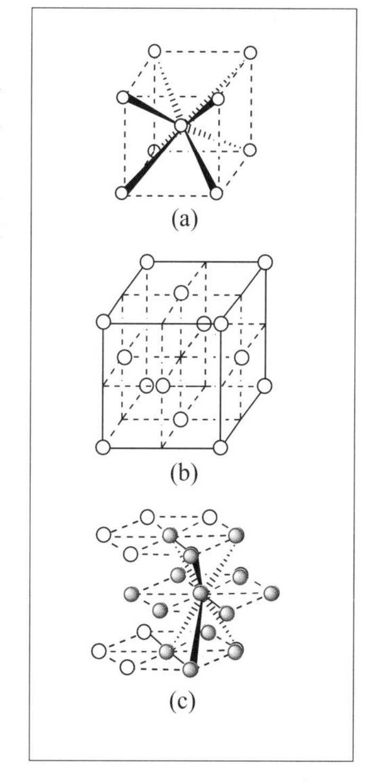

Figure 5.3 The three common metallic lattices: (a) body-centred cubic, (b) cubic closest packed and (c) hexagonal closest packed

Figure 5.3a is the **body centred cubic lattice (bcc)** in which the coordination number of each atom is eight.

In the bcc lattice there are six *next nearest neighbours* in the centres of the adjacent cubes only 15% further away, so that the coordination number may be considered to be 14.

The coordination number of any particular atom in a **cubic closest packed (ccp)** lattice is 12. Considering an atom in one of the face centres of the diagram in Figure 5.3b, there are four atoms at the corners of the face. Each face of the structure is shared between two such cells, and in the adjacent planes parallel to any shared face there are four atoms. These three sets of four atoms constitute the coordination number of 12.

The ccp structure is sometimes referred to as a **face-centred cubic (fcc)** lattice.

In the **hexagonal closest packed (hcp)** arrangement, shown in Figure 5.3c, the coordination consists of six atoms in the same plane as the atom under consideration plus three atoms from both of the adjacent planes, making a total of 12.

With such high coordination numbers it is quite clear that there can be no possibility of two-centre two-electron covalency because there are insufficient numbers of electrons. For example, metallic lithium has a body-centred cubic structure and coordination number of eight (or 14, if next nearest neighbours are considered). Each lithium atom has one valency electron and for each atom to participate in 14 or even eight electron-pair covalent bonds is not possible.

The high electrical conductivity of lithium (and metals in general) indicates considerable **electron mobility**. This is consistent with the molecular orbital treatment of an infinite three-dimensional array of atoms, in which the 2s orbitals are completely **delocalized** over the system with the formation of a band of $n/2$ bonding orbitals and $n/2$ anti-bonding orbitals for the n atoms concerned. Figure 5.4 shows a simple representation of the 2s band of lithium metal. The successive levels (molecular orbitals) are so close together that they almost form a continuum of energy. The formation of a band of delocalized molecular orbitals allows for the minimization of interelectronic repulsion. The very small gaps in energy between adjacent levels in a band of molecular orbitals allow a considerable number of the higher ones to be singly occupied. Such single occupation is important in explaining the electrical conduction typical of the metallic state. An electron may join the conduction band at any point of the solid and leave at any other point without having to surmount any significant energy barriers.

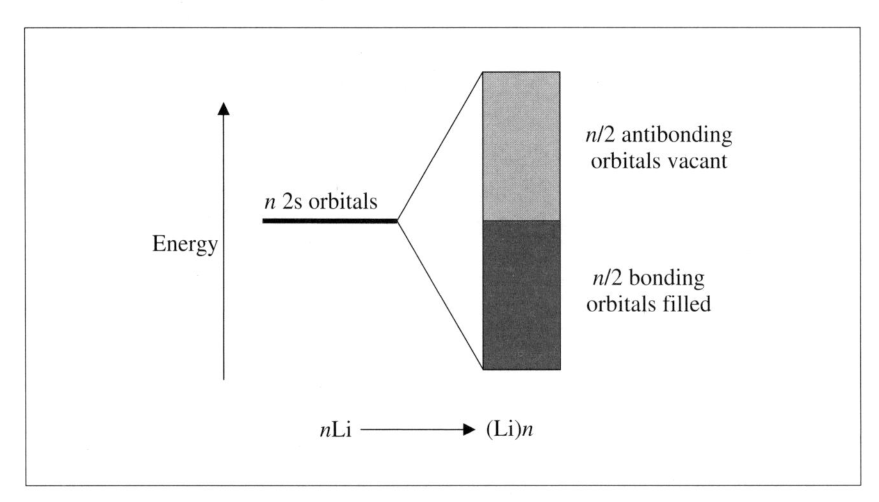

Figure 5.4 The 2s band of lithium metal

The band of molecular orbitals formed by the 2s orbitals of the lithium atoms, described above, is half filled by the available electrons. Metallic beryllium, with twice the number of electrons, might be expected to have a full "2s band". If that were so, the material would not exist, since the "anti-bonding" half of the band would be fully occupied. Metallic beryllium exists because the band of molecular orbitals produced from the 2p atomic orbitals overlaps (in terms of energy) the 2s band.

The overlapping of the 2s and 2p bands of beryllium, and their partial occupancies, are shown diagrammatically in Figure 5.5. The gap between the 2p and 2s bands of lithium is smaller than that in beryllium, so it is to be expected that some population of the 2p band in lithium occurs and contributes to the high conductivity of the metal. The dual population of the 2s and 2p bands in beryllium, with twice the number of electrons than in lithium, causes Be to have a higher conductivity than Li. The participation of the 2p band in the bonding of Be also explains the greater cohesiveness (bond strength) of the metal when compared to that of Li.

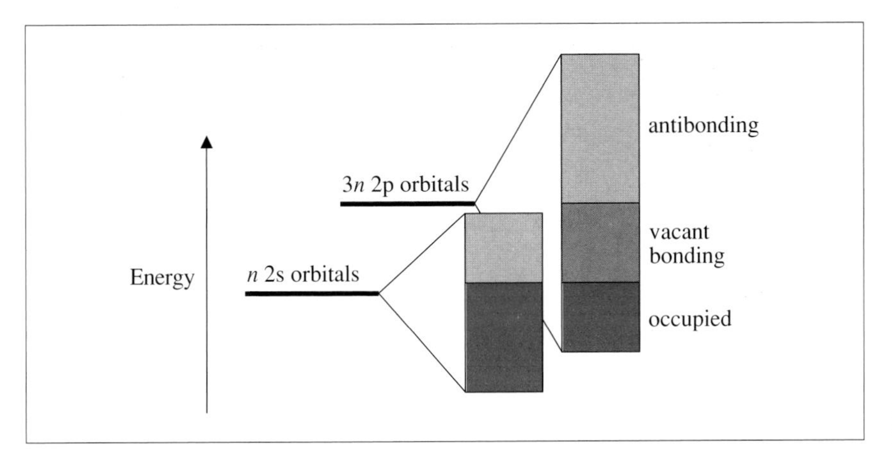

Figure 5.5 The 2s and 2p bands of beryllium

Worked Problem 5.1

Q What evidence is there to suggest that the bonding in beryllium is stronger than that in lithium?

A The melting and boiling temperatures of Be are 1280 °C and 2477 °C whereas those of Li are only 180 °C and 1330 °C. The enthalpies of atomization of Be and Li are 321 and 161 kJ mol^{-1}, respectively (this topic is dealt with in some detail in Chapter 6), again implying stronger bonding in Be.

If overlap between bands does not occur, the size of the band gap (between the lowest level of the vacant band and the highest level of the filled band) determines whether the element exhibits insulator properties or is a semiconductor. If the band gap is large, the material is an insulator. If the band gap is small, the material may be a semiconductor. Electrons may be transferred into the upper band thermally (semiconductivity increases with increasing temperature), by absorbing photons of suitable frequencies (photoconductivity) or by having impurities present which bridge the gap with their orbitals or bands (extrinsic semiconduction).

5.1.3 Van der Waals Forces

Van der Waals forces are described in Section 4.3.1. They are intermolecular non-valence forces which operate between molecules in their condensed phases and are particularly important in the Group 18 solids.

Worked Problem 5.2

Q What physical evidence is there for the differences in magnitude of valence and non-valence forces?

A Some physical data for the elements H, He, Ne and Na are tabulated below.

Element	*RAM/ RMM*	*m.p.*/°C	*b.p.*/°C	ΔH/kJ mol^{-1}		
				Fusion	*Evaporation*	*Atomization*
H_2	2	–259	–252	0.118	0.9	436
He	4	–270	–269	0.02	0.084	0
Ne	10	–249	–246	0.33	1.8	0
Na	11	97.8	890	2.6	101	109

The thermodynamic data are quoted in terms of enthalpy change per mole of the element as it exists in the condensed phases. The data for He and Ne show that the only cohesive forces operating are the non-valence intermolecular (interatomic in these cases) van der Waals forces, and so He and Ne have very low values of the physical constants and require only small amounts of energy to cause them to melt and boil. The dihydrogen molecule is subject to intermolecular forces only and requires only a small amount of energy to cause the element to melt or boil, but the covalent bond between the two hydrogen atoms is relatively hard to break as indicated by the enthalpy of atomization. Sodium metal is held together in the solid mainly by metallic bonding, with a very small contribution from intermolecular forces. It requires much more energy for melting than the other elements quoted, because some relaxation of the metallic bonding occurs in the process. It requires very much more energy to cause evaporation because that process occurs only when the metallic bonding is completely destroyed. A comparison of the data for H_2 and Na is some indication of the relative strengths of a single covalent bond and the metallic bonding produced by one valence electron per atom. The physical constants for Ne and H_2 indicate that the intermolecular forces in Ne are stronger than in H_2. In general, van der Waals forces are stronger between species with more electrons and with higher RMM values, the two being highly correlated.

5.1.4 Generalizations; Bonding in the Elements

For an element to have stable covalent molecules or extended arrays, it is essential that it has sufficient valence electrons and that these should experience a high effective nuclear charge, thus making the electron attachment energy sufficiently negative. Elements with electronegativity coefficients greater than about 1.8 fall into this class. Elements with electronegativity coefficients less than 1.8 do not have effective nuclear charges that are sufficiently high to attract electrons from other atoms in order to form covalent bonds, and they can achieve thermodynamic stability only by forming delocalized band structures characteristic of metallic bonding. As with any generalization based on an estimate of a property which is difficult to define, such as electronegativity, there are glaring exceptions. For example, silicon (χ = 1.7) and boron (χ = 2.0) are astride the 1.8 suggested borderline between metallic and non-metallic properties, but both are strictly non-metallic from the viewpoint of their not showing **metallic conduction** of electricity, *i.e.* where the conductivity is high and decreases with increasing temperature. Copper (χ = 1.8) and nickel (χ = 1.8) are on the borderline, but are certainly metals, copper having the next highest electrical conductivity to that of silver, which is the best conductor. Boron, silicon, germanium, arsenic, antimony, selenium and tellurium are sometimes described as **metalloid elements** or **semi-metals**. They show semiconductor properties, and have low electrical conductivities which increase as the temperature increases.

The non-metallic elements whose cohesiveness depends upon participation in strong covalent bond formation exist in the following forms:

(i) **Diatomic molecules** which may be singly bonded, *e.g.* H_2 and F_2; doubly bonded, *e.g.* O_2; or triply bonded, *e.g.* N_2.

(ii) **Small polyatomic molecules** involving single covalent bonds between adjacent atoms, *e.g.* P_4 and S_8. The structures of the P_4 and S_8 molecules are shown in Figure 5.6.

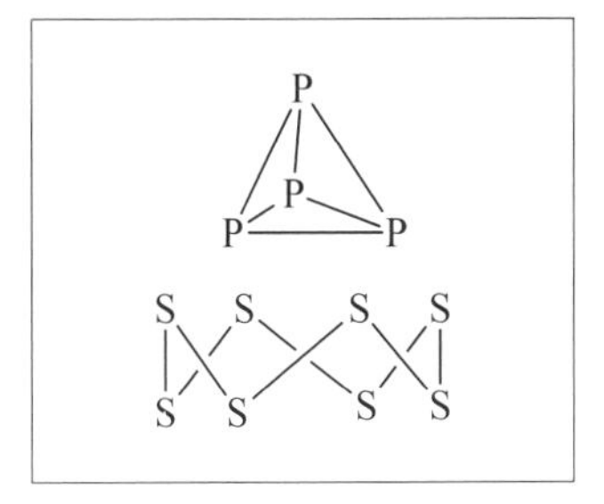

Figure 5.6 The structures of the P_4 and S_8 molecules

(iii) **Long chains** which are polymers of small molecules, *e.g.* "plastic" sulfur composed of polymers of S_8 units with variable chain lengths.

(iv) **Sheets** of atoms held together by van der Waals forces, as in forms of black phosphorus and graphite.

(v) **Three-dimensional infinite arrays**, *e.g.* diamond, boron and silicon.

5.1.5 Ionic Bonding and the Transition to Covalency

Because subsequent chapters deal with *compounds* in which the chemical bonding can be either covalent or ionic, a brief description of ionic bonding is included here and a discussion of the atomic properties which govern the transition from ionic to covalent character. As indicated in

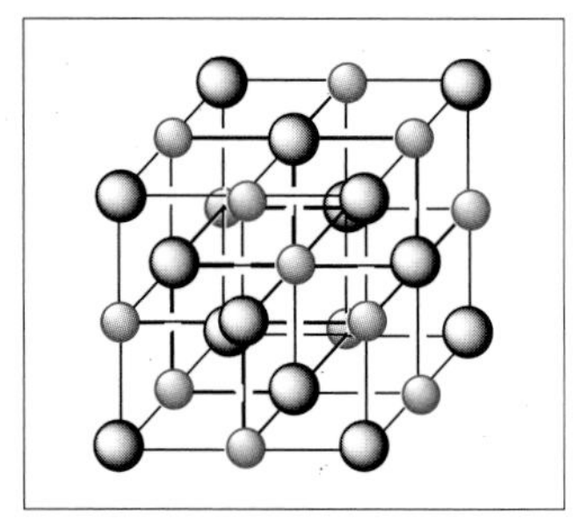

Figure 5.7 The crystal structure of NaCl

In a crystal lattice, the +/– attractive forces outweigh the +/+ and –/– repulsive forces because the ions of opposite charges are closer to each other than the ions of like charges.

Section 4.4.4, when a binary compound is formed between elements with very different values of their electronegativity coefficients, ionic bonding results. The classical example of ionic bond formation is that in sodium chloride. The crystal structure of NaCl is shown in Figure 5.7.

Each sodium ion is surrounded by six chloride ions and each chloride ion is surrounded by six sodium ions. The **coordination numbers** of both ions are six. In the formation of NaCl, the sodium atom loses its valency electron ($3s^1$) to the chlorine atom ($3s^23p^5$) so that Na^+ and Cl^- have the electronic configurations of neon and argon, respectively, in accordance with the octet rule (discussed in Section 5.3). Because the electron attachment enthalpy of chlorine (–355 kJ mol^{-1}) does not provide compensation for the enthalpy used to ionize the sodium atom (500 kJ mol^{-1}), the stability of the compound is produced by the attractive forces operating between the oppositely charged ions in a *crystal lattice*.

When the lattice is formed from the gaseous ions the **lattice enthalpy** of NaCl is released, *i.e.* $\Delta H^{\ominus} = -784$ kJ mol^{-1}, thus overcoming the imbalance of the ionization processes to make the compound thermodynamically stable. There is a significant, but relatively small, contribution from van der Waals forces to the lattice enthalpy. The enthalpy changes accompanying the formation of NaCl from its elements are shown in Figure 5.8. To transform the elements into their gaseous atomic states the enthalpies of atomization are included in the formation process.

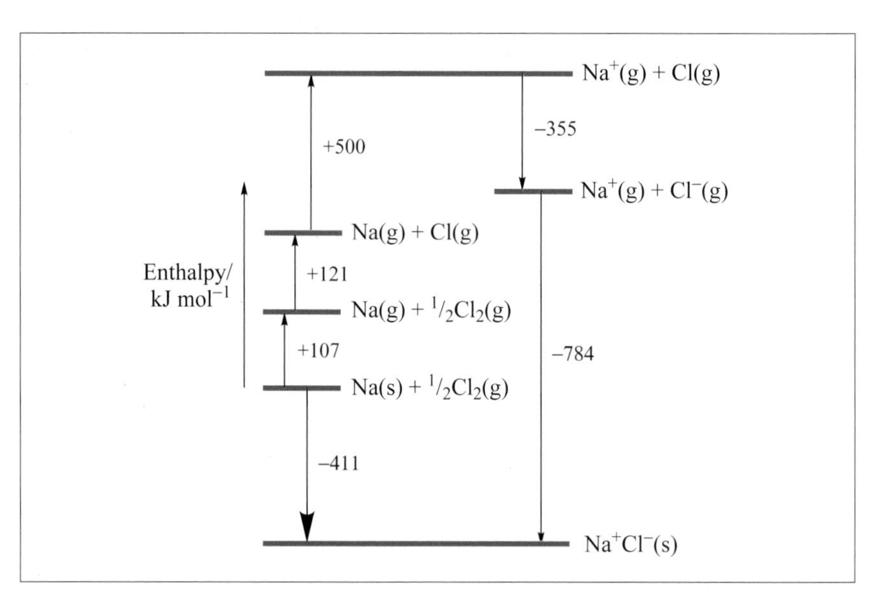

Figure 5.8 The enthalpy changes accompanying the formation of NaCl from its elements

From Figure 5.8 it may be concluded that the factors which favour the formation of an ionic compound, *i.e.* that make the enthalpy of formation more negative, are:

(i) small atomization enthalpies of the combining elements;

(ii) a low ionization enthalpy of the electropositive element;

(iii) a large negative value of the electron attachment enthalpy of the electronegative element; and

(iv) a large negative lattice enthalpy.

Any deviations from these criteria result in compounds which are less ionic, more covalent.

5.1.6 Character of Chemical Bonds: Ionic, Covalent or Both?

Both ionic bonding and covalent bonding are extreme descriptions of the real state of affairs in elements and their compounds, and although there are many examples of almost pure, *i.e.* 99%, covalent bonding, there are very few so-called ionic compounds in which there is complete transfer of one or more electrons from the cationic element to the anionic one. This ionic–covalent transition in which an initially purely ionic compound acquires some covalent character is reasonably easy to deal with qualitatively, but is very difficult to quantify. Fajans stated four rules which are helpful in deciding which of two compounds has the more covalency. In this respect the starting point is complete electron transfer in the compounds considered. The positively charged cations are then considered to exert a polarizing effect on the negatively charged anions such that any polarization represents a clawing-back of the transferred electron(s), so that there is some electron sharing between the atoms concerned which may be interpreted as partial covalency. Figure 5.9 shows diagrammatically what is meant by polarization of an anion; the distortion from the spherical distribution of the anionic electrons is a representation of the covalency.

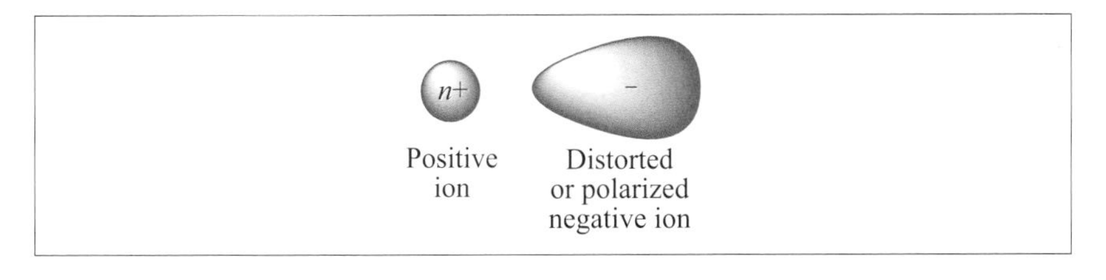

Figure 5.9 The polarization of an anion by a cation

Fajans' rules state that the extent of polarization of an anion or the partial covalent character of an otherwise 100% ionic compound is increased by:

1. A high charge of the cation and/or anion.
2. A small cation.
3. A large anion.
4. A cation with a non-inert-gas configuration.

The melting temperatures of NaCl, $MgCl_2$ are 801 °C and 714 °C, respectively, and Al_2Cl_6, which is a molecular Al–Cl–Al doubly bridged compound in its anhydrous state, sublimes at 180 °C.

The reasoning underlying these rules is dependent upon the **charge density** of the cation, the charge density being the ratio of the charge on the ion to its surface area. The first three rules are straightforward electrostatics. Comparing the anhydrous chlorides of sodium, magnesium and aluminium it is expected that the last compound would have the greatest partial covalency, be the least ionic, and that the sodium compound would be the most ionic.

The fourth rule is an indication that a sub-shell of d electrons is a poorer shield of the nuclear charge than an octet of s^2p^6 electrons, and that a cation with a d shield will exert a greater polarizing effect than one with an s–p shield. Comparing copper(I) chloride with sodium chloride is the classical example of this rule. Cu^I has the electronic configuration $3d^{10}$ whereas Na^+ has the [Ne] core configuration. The ions are similar in size, their ionic radii being 96 pm (Cu^+) and 102 pm (Na^+), but the melting temperatures of their chlorides are 430 °C (CuCl) and 801 °C (NaCl), an indication that the copper compound is more covalent.

Worked Problem 5.3

Q Show that Fajans' rules are consistent with the criteria for ionic compound formation described in Section 5.1.5.

A High charges on the cation and anion contribute to a high lattice energy, but would require large amounts of enthalpy in their formation, and would encourage polarization to promote covalency. A small cation would be associated with a relatively high value of the ionization energy, and would promote covalency. A large anion would be associated with a relatively low negative value for the enthalpy of electron attachment and would favour covalency.

Worked Problem 5.4

Q In the supposedly ionic compounds MX and NX_2, where M and N are metallic elements and X is a univalent non-metal, which compound would you expect to have a greater degree of covalency?

A The element N would have a double charge in the compound NX_2 which would give it a higher polarizing power than the singly charged element M in the compound MX. The compound NX_2 would have a greater degree of covalency.

5.2 Valency and Oxidation State: Differences of Terminology

Valency is possibly the most used and the least well-defined term in theoretical chemistry! The valency of an element is its *combining power* represented as the number of electron-pair bonds in which it can participate by sharing some or all of its valency electrons with an *equal number* from the atoms to which it is bonded. This number is equal to or sometimes less than the number of valency electrons, *i.e.* those electrons which are capable of entering into covalent bonding, the electrons in the valence shell. These are the "outer" electronic configuration of the atom, and many examples are given in this and subsequent chapters. For example, the sulfur atom has an outer electronic configuration $3s^2 3p^4$ and could be expected to be six-valent if all the valence electrons took part in covalent bonding, as in SF_6. In such a case the central atom is said to be covalently saturated. Sulfur also exerts valencies which are less than the maximum, as in the compounds SF_4 (four-valent S), SCl_2 (two-valent S) and H_2S (two-valent S).

Ionic compounds have indications of the "valency" of the constituent elements in their stoichiometries, *i.e.* in NaCl both atoms are monovalent, in MgF_2 the magnesium is divalent in combining with the monovalent fluorine. In such compounds the term "valency" has to be used with caution. In NaCl each ion has a coordination number of six, *i.e.* each ion has six nearest neighbours which might be regarded as being ionically bonded to the central ion. In the structure of MgF_2 (the rutile structure of TiO_2), each magnesium ion is surrounded by six fluoride ions, and each fluoride ion is surrounded by three magnesium ions arranged in a trigonal plane. The terms monovalent and divalent, for Na and Mg respectively, are used to describe the metals in the compounds mentioned, but the oxidation state concept is more accurate and understandable.

There are important differences between the two concepts which should be appreciated. Oxidation state is a *formal charge* on an atom which is in combination with one of the very electronegative atoms, *i.e.* F or O, in which their oxidation states are deemed to be –1 and –2, respectively. For example, in MgF_2 and MgO the oxidation state of the magnesium is +2 (*i.e.* II in the *Roman numerals* conventionally used to indicate oxidation states: Mg^{II}) and equal to the charge on the Mg^{2+} ions in those compounds. In ionic compounds there are no electron-pair bonds so the strict definition of valency, given above, does not apply.

In essentially covalent compounds the oxidation state concept can be useful in their classification. The two main oxides of sulfur are SO_2 and SO_3 and, if the oxidation state of oxygen is taken to be –2 by convention, the oxidation states of the sulfur atoms in the oxides are +4 and +6, respectively, *i.e.* they contain S^{IV} (sulfur-4) and S^{VI} (sulfur-6).

Worked Problem 5.5

Q Explain why the oxidation state of oxygen in the oxide ion is taken to be –2.

A In fully ionic oxides the oxygen species present is O^{2-}; therefore the oxidation state of the oxygen is equal to the charge, *i.e.* –2. This value is taken to be the oxidation state of oxygen in compounds that are not necessarily ionic.

The oxidation states of sulfur in the compounds, SF_4 and SCl_2, +4 and +2 respectively, may be included in their formulae as right-hand superscripts in Roman numerals: $S^{IV}F_4$ and $S^{II}Cl_2$. This convention is used whenever the oxidation state of an element needs to be emphasized. For example, the compound Fe_3O_4 may be regarded as a mixed oxide, $Fe_2O_3.FeO$, and could be written as $Fe^{III}{}_2Fe^{II}O_4$. Negative oxidation states are rarely indicated in formulae. The phrases "in SF_4 the sulfur atom is in oxidation state IV" and "in SF_4 the sulfur atom is in its +4 oxidation state" should be regarded as identical in meaning.

The sulfur atoms in the compounds SF_4, SCl_2 and H_2S have valencies of 4, 2 and 2, but if the formal oxidation states of the fluorine, chlorine and hydrogen atoms are taken to be –1, –1 and +1, respectively, the oxidation states of the sulfur atoms in those compounds are +4, +2 and –2 respectively.

In H_2S the hydrogen atom is assigned the oxidation state of +1 because hydrogen is less electronegative than sulfur. In a compound such as LiH (which is ionic, Li^+H^-) the hydrogen is in oxidation state –1, consistent with H being more electronegative than Li. Similarly, but contrarily, the nitrogen atom in the ammonia molecule, NH_3, is considered to be in the –3 oxidation state because it may be produced by reducing elementary nitrogen by dihydrogen. In the chlorate(VII) ion (perchlorate), ClO_4^-, the chlorine atom has an oxidation state of +7 because the four oxygen atoms are considered to be in the –2 state, and a central chlorine(VII) makes the overall charge minus one. The bonding in the ion is essentially covalent, and the massive amount of energy which would have to be used to remove (ionize) the seven valency electrons of the chlorine atom is sufficient to prevent any ionic bonding between the chlorine atom and the four oxygen atoms.

Ionization of any transition element brings about a resolution of the problem of the closeness of the *n*s electrons and the (*n* – 1)d electrons in favour of the d electrons, as they are sufficiently lower in energy than the s.

The classical cases of distinction between valency and oxidation state are in the **coordination complexes** of the transition elements. For example, in the **complex compound** $[Cr(NH_3)_6{}^{3+}](Cl^-)_3$ the **complex ion** containing the chromium ion, $Cr(NH_3)_6{}^{3+}$, has a chromium atom at its centre which engages in six electron-pair bonds with the six ammonia ligand molecules, the latter supplying two electrons each in the formation of six coordinate or dative bonds. To that extent the chromium atom is participating in six electron-pair bonds and could be called six-valent. Because the chromium atom is using none of its valency electrons, to think of it as exerting any valency is misleading. What is certain is that the complex ion exists as a separate entity with three chloride ions as counter ions, and therefore has an overall charge of +3. Since the ammonia ligands are neutral molecules, the chromium atom may be thought of as losing three of its valency electrons to give the +3 oxidation state, written in the form Cr^{III}. The original electronic configuration of the chromium atom's valency shell is $4s^1 3d^5$, which in its +3 state would become $3d^3$. The complex ion is fully described as the hexaamminechromium(III) ion. When discussing complexes of the kind described, it is useful to know that the metal ion has a particular d configuration.

The atoms of elements in their standard states have, by convention, oxidation states of zero, although they mostly participate in either metallic or covalent bonding and are thus exerting their valencies, although in the case of metals there are no conventional electron-pair bonds.

Worked Problem 5.6

Q Elementary sulfur consists of S_8 molecules arranged in a puckered ring. What is the valency of each sulfur atom and what is the oxidation state of the element?

A Each sulfur atom forms single bonds with its two neighbours in the ring. The valency of the sulfur atoms is 2. The oxidation state of the element is zero (by convention, which applies to all elements).

5.3 Valency and the Octet and 18-Electron Rules

There are two rules which govern many of the valencies or oxidation states of the elements in their compounds. One is the **octet rule**, which applies to the elements of the 2nd and 3rd periods and to the s- and p-block elements generally; the other is the **18-electron rule**, which applies to the elements of subsequent periods, in particular to the compounds formed by the transition elements. The non-metallic p-block elements of the periods beyond the third are subject to the 18-electron rule, but since they all possess a filled d^{10} shell, they may be considered to obey the octet rule in some of their compounds.

5.3.1 The Octet Rule

The octet rule was stated by G. N. Lewis in 1923:

"Atoms in combination have a tendency to achieve an octet of electrons or a share in an octet."

In modern terms, this means that atoms with the electronic configurations $s^n p^m$ (varying from n = 1, m = 0 to n = 2, m =5) either (i) lose the *ns* and *m*p electrons to give a positive ion with the electronic configuration of the previous Group 18 element, (ii) gain a sufficient number of electrons to give a negative ion with the s^2p^6 configuration of the next Group 18 element, or (iii) share an s^2p^6 configuration with the atom(s) with which it is combined. The various possibilities are given in Table 5.1 for positive or negative ion formation and the stoichiometry of possible hydrides of an element, E.

Table 5.1 Application of the octet rule to elements of the s- and p-blocks

Hydride	*n*	*m*	*Positive ions formed*	*Negative ions formed*
EH	1	0	$E^{(n+m)+}$	None formed
EH_2	2	0	$E^{(n+m)+}$	None formed
EH_3	2	1	$E^{(n+m)+}$	None formed
EH_4	2	2	$E^{(n+m)+}$	$E^{[8-(n+m)]-}$
EH_3	2	3	None formed	$E^{[8-(n+m)]-}$
EH_2	2	4	None formed	$E^{[8-(n+m)]-}$
EH	2	5	None formed	$E^{[8-(n+m)]-}$
None formed	2	6	Zero-valent	

Worked Problem 5.7

Q Give examples of Group 14 compounds containing E^{4+} or E^{4-} ions. Why are there few examples of these kinds of compound?

A PbO_2 crystallizes in the rutile (TiO_2) form and has Pb^{4+} ions surrounded by six oxide ions. The compound Al_4C_3 contains Al^{3+} ions and C^{4-} ions in a complicated structure, with the distance of closest approach of carbon ions of 316 pm, meaning that there is very little possibility of any covalency between them. The compound reacts with water to give methane, CH_4, and has been described as a **methanide**. As indicated by Fajans' rules, there is a high probability that any compounds containing ions with charges of either 4+ or 4– would have a high degree of covalency.

The octet rule as detailed in Table 5.1 accounts for many of the ions and compounds formed by the elements of the s- and p-blocks of the Periodic Table. The basis of the rule is that any *ions* formed are likely to have the largest positive or negative charges possible, because these offer the greatest electrostatic attraction leading to the greatest stability for the compound produced. The difficulty of forming positive ions increases across any period, because successive ionization energies increase with the number of electrons removed. Removal of the valency electrons is relatively easy compared to removing those from the next lower shell of electrons, so the limits of positive ion formation are determined by the number of valency electrons. For the elements of Groups 15–17, which are high up on the electronegativity scale, it is an energetically better option for them to form negative ions, but the formulae of these are determined by the octet rule in that they tend to accept electrons only so that the relevant p orbitals are completely filled and the electronic configuration is that of the Group 18 element of the particular period. No further additions of electrons would give an energetic advantage, as they would have to occupy the next available s orbital of the next higher energy shell. Successive electron attachment energies are increasingly endothermic because of repulsion, so the attachment of more than one electron, although endothermic, can only occur if the resulting compound has a sufficiently large lattice energy to compensate for the endothermicity.

The many covalent compounds which are formed between elements of the s- and p-blocks have stoichiometries that are determined by the sharing of the valence electrons of the central atom such that there are eight electrons shared between the central atom and its ligand atoms. The expected valency is that consistent with the four valence orbitals, one s + three p, and is therefore $n + m$ for values of $n + m \leq 4$, and $8 - (n + m)$ for values of $n + m \geq 4$, where n is the number of s electrons and m is the number of p electrons in the valence shell of the central atom. The normal outcome of combination between two elements is the maximization of the number of covalent bonds with respect to the above formula. If more or fewer bonds are formed, such cases are exceptions to the rule.

Although no positive ions are formed with a charge greater than 4+, there are many compounds of the elements of Groups 15–17 that have stoichiometries which indicate that the element has been oxidized to a greater degree than that expected by the octet rule. For example, NO_3^-, PCl_5, SF_6 and IF_7 are species in which the valencies of the central elements are 5, 5, 6 and 7, respectively. The oxidation states of those elements are +5, +5, +6 and +7, respectively, but there is a problem with NO_3^- which does not apply to the other examples. The central nitrogen atom has only the 2s and 2p orbitals in its valence shell, and the next orbital of lowest energy is the 3s which is very unlikely to be used because of the 3s–2p energy gap. Nitrogen chemistry must be explicable in terms of the *four* atomic orbitals that can contribute to its bonding as a central atom. In terms of electron pair bonds, the nitrate ion can be written as in Figure 5.10, which shows the nitrogen atom participating in a σ bond to the oxygen atom with a single charge, a double (σ + π) bond

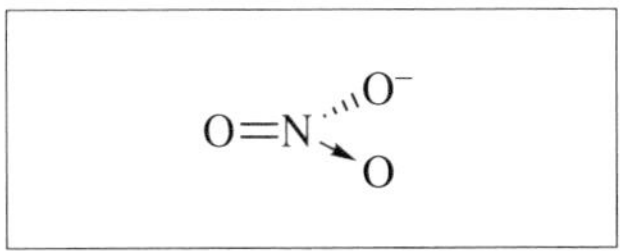

Figure 5.10 A structure of the nitrate(V) ion

to one of the other oxygen atoms, and making a coordinate or dative bond to the third oxygen atom.

Such a formulation allows the nitrogen to be bonded to the three oxygen atoms by using its 2s and 2p electrons, but shows the difficulty that arises with both valency and oxidation state concepts. The formula allows the nitrogen atom to exhibit its expected valency of three in that it participates in three electron pair bonds, not counting the coordinate bond as such. The oxidation state of the nitrogen atom is +5 if the oxygen atoms are counted as –2: $N^{V} + 3O^{-2}$.

Some of the other exceptions to the rule, mentioned above, are examples of **hypervalency**. This occurs when the valence shell expands to include the next lowest energy orbitals, so that more covalent bonds can be formed than the number expected from the octet rule. In PCl_5 the $3s^23p^3$ electronic configuration of the phosphorus atom is expected to be trivalent because of the three unpaired electrons in the 3p orbitals. An **expansion of the valence shell** in P is possible if the two 3s electrons are uncoupled, and one is promoted to one of the 3d orbitals to give the configuration $3s^13p^33d^1$, which would allow the P atom to be five-valent. In the case of SF_6, the sulfur atom has expanded its valence shell to give the configuration $3s^13p^33d^2$, so becoming six-valent. Likewise in IF_7 the central iodine atom has expanded its valence shell to give the configuration $5s^15p^35d^3$, to become seven-valent. In the compounds PCl_5, SF_6 and IF_7 the central atoms have the oxidation states indicated by P^{V}, S^{VI} and I^{VII}.

In extreme exception to the octet rule, xenon forms, for example, the compounds XeF_2, XeF_4, XeF_6, XeO_3 and XeO_4. In terms of the expansion of its valence shell this means the promotion of a 5p electron to a 5d orbital to form the divalent state of Xe, the promotion of two 5 p electrons to 5d orbitals to form the four-valent state, the promotion of three 5p electrons to 5d orbitals to give the six-valent state, and the extra uncoupling and promotion of a 5s electron to a fourth 5d orbital to form the eight-valent state. Such electronic juggling does explain the bonding in these exceptional compounds in terms of electron-pair bonds, but there is an alternative approach via molecular orbital theory which does not require these promotions in order to explain the bonding in hypervalent compounds. This is left to more specialist books, except to remark that all hypervalent compounds have bonds to the central atom which are weaker than those in compounds which are consistent with the octet theory, implying that the bonds are not truly electron-pair bonds of the conventional kind.

There are other exceptions to the octet rule, exemplified by the compounds of the heavier p-block elements, *e.g.* Tl^+, $SnCl_2$, $PbCl_2$ and $BiCl_3$, which possess oxidation states/valencies that are two units less than expected from the rule. This phenomenon is called the **inert-pair effect**.

It is called so because the s^2 pair of valency electrons do not participate in ion formation, nor do they enter into covalent bonding. The oxidation states of these positive ions are lower by two units than expected from their electronic configuration, as are the numbers of single bonds entered into by the atoms in their compounds. The effect is explained by relativistic effects on the valency s and p orbitals, as discussed in Section 4.5.

5.3.2 Compound Stoichiometry (s- and p-Block Compounds)

The stoichiometries of the ionic compounds of the s- and p-block metals are largely determined by the factors responsible for the magnitude of the **standard enthalpy of formation** of such compounds. The factors which make the formation of an ionic compound favourable, *i.e.* which contribute to making the enthalpy of formation as negative as possible, are:

(i) A low atomization enthalpy of the metal.
(ii) A low atomization enthalpy of the non-metal.
(iii) Low ionization enthalpy or enthalpies of the metal atom.
(iv) A highly negative electron attachment enthalpy of the non-metal atom.
(v) A highly negative lattice enthalpy for the crystal structure as formed from the gaseous ions.

The condition for the formation of a thermodynamically stable ionic compound may be written in the form:

$$-\text{Lattice enthalpy} > \Sigma(\text{ionization enthalpy}) + \Sigma(\text{electron attachment enthalpy}) + \Sigma\Delta_a H^\ominus$$

which, if fulfilled, makes the enthalpy of formation of the compound negative.

The dominant factors are (iii) and (v), which are in opposition to each other. In order to achieve a high negative lattice enthalpy a high cation charge is required, and that can only be provided if more than one electron is removed in the ionization stage. Successive ionization enthalpies increase as more electrons are removed, but there comes a stage where the next electron to be removed is from a shell in the electron core which is so energy-expensive that the larger negative lattice enthalpy cannot compete, so the compound does not exceed the stoichiometry appropriate to the removal of all the valence electrons of the metal atom. Thus the ionic compounds of the s- and p-block metals all show the typical Group oxidation state, *i.e.* $n + m$ for values of $n + m \leq 4$ and $8 - (n + m)$ for values of $n + m \geq 4$, where n is the number of s electrons and m is the number of p electrons in the valence shell of the central atom.

For example, Group 14 elements have the valence shell configuration s^2p^2, which has two unpaired electrons which could be used to make electron-pair bonds with two uni-valent fluorine atoms, the pair of s electrons remaining as they were in the separate atom. In such a case the compound MF_2 would be produced. If energy were to be expended in unpairing the two s electrons and promoting one of them to the otherwise vacant third p orbital, a four-valent state would be created with the configuration sp^3 with all four valency electrons occupying single orbitals ready to pair up with electrons from four fluorine atoms to give the compound MF_4.

This is not the case at the right-hand side of the Periodic Table, where the differences between electronegativity coefficients are small and allow covalent bonds to be formed rather than ionic ones in the binary compounds of those elements. In most cases of covalency there are compounds formed which exhibit the group valency and some which have valencies that are less than and more than the group valency. Additionally it is observed that the valencies of the central elements usually vary by two units. These observations are rationalized by the need to unpair electrons to increase the valency of an atom, and that the unpairing allows the valency of the atom to increase by two units.

5.3.3 The 18-Electron Rule

There are many exceptions to the 18-electron rule. Only a few examples are given here because the rule is relevant to transition metal *compounds* rather than to the elements themselves.

The valence shells of the transition elements, to which the 18-electron rule applies, contain s, p and d orbitals that altogether can accommodate 18 electrons; hence the name of the rule. The 18-electron rule is that there is a tendency for the central atoms in compounds of the transition elements to possess a share in 18 electrons.

Towards the middle and beyond of the transition series there is reasonable adherence to the rule, *e.g.* in the neutral compound $Fe(CO)_5$ the iron atom has eight valency electrons in its zero oxidation state and the five CO ligand molecules each supply two electrons in the formation of coordinate bonds, making $8 + (2 \times 6) = 18$ electrons in the Fe valence shell. In the complex ion $[Fe(CN)_6]^{4-}$, iron is in its +2 oxidation state, *i.e.* the overall charge of –4 is produced by the six negative CN^- ions offset by a formal charge of +2 on the iron centre. There are then six of the iron atom's valency electrons in Fe^{II}, and these are joined by the 12 electrons, two from each coordinated cyanide ion, to make up the required 18. This is not so in the $[Fe(CN)_6]^{3-}$ complex ion, which has only 17 electrons in the valence shell of the iron atom. There are other cases where the valence shells of central atoms have fewer than 18 electrons, but the rule holds generally with the organometallic compounds of the transition elements, *e.g.* in bis(benzene)chromium(0), $Cr(C_6H_6)_2$ (Figure 5.11), the chromium atom is in its zero oxidation state (6 valency electrons) and the two benzene molecules supply their complement of 6π electrons each (*i.e.* 12 electrons) to make up the required 18.

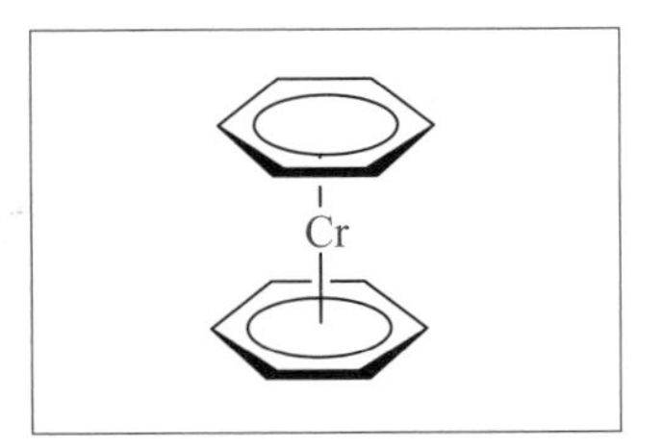

Figure 5.11 The structure of bis(benzene)chromium

5.3.4 Compound Stoichiometry (d-Block Compounds)

The main difference between the stoichiometries of compounds of the d-block elements and those of the s- and p-blocks is that d-block elements commonly differ in oxidation state by single units whereas elements of the s- and p-blocks usually differ by two units, *e.g.* chromium forms the fluorides CrF_2, CrF_3, CrF_4 and CrF_5, but sulfur forms SF_2, SF_4 and SF_6. The principle reason for this difference is that electron pairs of the s- and p-block elements have to be uncoupled to increase their valencies/oxidation states, but there is no uncoupling necessary with the earlier d-block elements which contain unpaired electrons in their valence shells, some or all of which are used when the element forms compounds. This topic is extended in Section 7.2.4.

5.4 Periodicity of Valency and Oxidation States in the s- and p-Block Elements

5.4.1 Hydrogen

The first element, hydrogen, forms binary compounds with many of the elements and does so in three main ways.

(i) By ionization to give the proton, H^+, which exists in aqueous solutions of acids as the **hydroxonium** or **oxonium ion**, $H_3O^+(aq)$, in which the proton is covalently bonded to the oxygen atom of a water molecule, and the resulting ion is **hydrated** (surrounded by a number of other water molecules loosely bonded by hydrogen bonding interactions). Perhaps the best description of the bonding is to liken H_3O^+ to the ammonia molecule NH_3: the two species are isoelectronic, and O^+ has the same electronic configuration as N, $2s^2 2p^3$.

In this book, and most others, the proton as a chemical entity in aqueous solution is represented as $H^+(aq)$, $H_3O^+(aq)$ or just H^+ when it is clear that solutions are being discussed.

(ii) By accepting an electron to form the hydride ion, H^-. Ionic or **saline hydrides** are formed by the elements of Group 1, the heavier elements of Group 2, Ca, Sr and Ba, and by the lanthanides. They all react with water to give dihydrogen and a solution of the metal hydroxide:

$$MH(s) + H_2O(l) \rightarrow M^+(aq) + OH^-(aq) + \frac{1}{2}H_2(g)$$

(iii) By entering into covalent bond formation using its single valence electron, *e.g.* HF, H_2O.

In addition to the three types of chemical bonding, hydrogen is essential for all those intermolecular interactions known as hydrogen bonding.

5.4.2 The s- and p-Block Elements other than Hydrogen

Table 5.2 shows the variations of valency, *i.e.* the number of single electron-pair bonds, of the elements of the s- and p-blocks. The oxidation states (Roman numerals) of the metallic elements are indicated instead of their valency. The elements of the 2nd period, Li to Ne, show the values expected from the strict application of the octet rule. The elements of the subsequent periods follow the rule, but there are many exceptions

The most spectacular variations from the octet rule are in xenon chemistry, where the rule would predict zero valency. Xenon forms a small range of compounds with oxygen and fluorine in which it expands its valence shell to produce valencies of 2, 4, 6 and 8.

Table 5.2 The valencies (Arabic numerals)/oxidation states (Roman numerals) of the elements of the elements of the s- and p-blocks; the valencies shown in **bold red** are hypervalencies, those in *italic red* are the result of the inert pair effect

Group							
1	*2*	*13*	*14*	*15*	*16*	*17*	*18*
Li	*Be*	*B*	*C*	*N*	*O*	*F*	*Ne*
			4				
		3		3			
	II				2		
I						1	
							0
Na	*Mg*	*Al*	*Si*	*P*	*S*	*Cl*	*Ar*
						7	
					6		
				5		**5**	
			4		**4**		
		III		3		**3**	
	II		*2*		2		
I		*1*				1	
							0
K	*Ca*	*Ga*	*Ge*	*As*	*Se*	*Br*	*Kr*
						7	
					6		
				5		**5**	
			4		**4**		
		III		3		**3**	
	II		*2*		2		**2**
I		*1*				1	
Rh	*Sr*	*In*	*Sn*	*Sb*	*Te*	*I*	*Xe*
							8
						7	
					6		**6**
				5		**5**	
			4		**4**		**4**
		III		3		**3**	
	II		*2*		2		**2**
I		*I*				1	
Cs	*Ba*	*Tl*	*Pb*	*Bi*	*Po*	*At*	*Rn*
							6?
				5			
			4		**4**		
		III		3			
	II		*II*		2		**2**
I		*I*				1	

because of hypervalency (because low-lying d orbitals may be used to expand the valence shell) and the onset of the inert pair effect.

5.4.3 Negative Oxidation States of the s- and p-Block Elements

The electronegative elements of the p-block form ionic compounds with metals in which they have negative oxidation states.

A summary of the negative oxidation states of the s- and p-block elements, as they exist in monatomic ions, is given in Table 5.3. Examples of negative ions containing more than one s- or p-block element exist (*e.g.* O_2^{2-}, the peroxide ion), but are not detailed in this text.

Even the Group 1 elements can exist as uni-negative ions under certain conditions. For example, sodium metal dissolves in an ethoxyethane (diethyl ether) solution of the cryptand (*i.e.* a cage-like ligand) known as C222, the structure of which is shown in Figure 5.12.

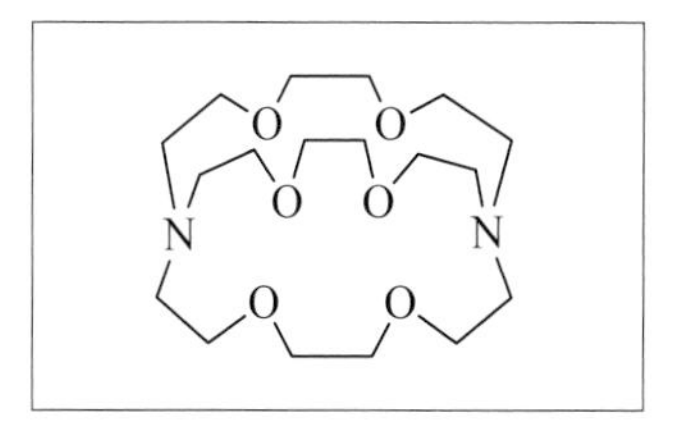

Figure 5.12 The structure of the C222 cryptand

The sodium metal disproportionates to give Na^+ cations which are stabilized by interaction with the cryptand molecule to give a cryptate ion, $Na[C222]^+$, leaving the other half of the sodium as the solvated Na^- anions. The solid compound $[Na(C222)]^+Na^-$ crystallizes from the solution, and is known as an **alkalide**.

Table 5.3 Negative oxidation states of the s- and p-block elements

Group 1	2	13	14	15	16	17
H						
−1						
Li	*Be*	*B*	*C*	*N*	*O*	*F*
−1			−4	−3	−2	−1
Na	*Mg*	*Al*	*Si*	*P*	*S*	*Cl*
−1			−4	−3	−2	−1
K	*Ca*	*Ga*	*Ge*	*As*	*Se*	*Br*
−1				−3	−2	−1
Rb	*Sr*	*In*	*Sn*	*Sb*	*Te*	*I*
−1				−3	−2	−1
Cs	*Ba*	*Tl*	*Pb*	*Bi*	*Po*	*At*
−1				−3	−2	−1

Apart from the Group 1 elements, the negative oxidation states are produced by the p-block elements and these are consistent with the octet rule. Their electronic configurations are those of the Group 18 element of the particular period.

5.5 Oxidation States of the Transition Elements

The oxidation state concept is best used for the transition elements, and Table 5.4 shows the oxidation states exhibited by the three series of transition elements in their many and varied compounds. The oxidation states shown in bold red are those that are the most stable in acidic aqueous solution in the presence of air. Any oxygen present would possibly oxidize an otherwise stable state, *e.g.* Fe^{2+} ions are the most stable state of iron in aqueous solution in the absence of dioxygen which would (slowly) oxidize the Fe^{2+} ions to Fe^{3+}. The states that are most stable in the absence of air are shown in italic red, if different. As the oxidation state of any element increases, there is an increasing tendency for the compounds of that element to be covalent. In low positive oxidation states the compounds are ionic and have high melting points, but as the oxidation state increases the compounds tend to be covalent and molecular and so have low melting points.

From left to right across the series of transition elements there is a triangular pattern shown by the *maximum* oxidation states observed, which is regular in the first series, equal to the number of s and d valence electrons up to the half-way stage in Group 7, and then decreasing regularly as pairing up of the valence electrons is brought about by the influence of the Pauli principle. The pattern in the second series is nearly regular, with two states of ruthenium (+7 and +8) and one state of rhodium (+6) with maxima greater than expected for use of the unpaired valence electrons. In the third series there are more deviations from regularity, with osmium (+7 and +8) and iridium (+6) showing the same extra oxidation states as Ru and Rh in the second series, and some extra deviations with platinum (+5 and +6) and gold (+5). In the *first half* of each series the maximum oxidation states correspond to the complete removal of the valence electrons of each element, *i.e.* the valence shell configurations are $s^2d^{\text{group number} - 2}$, $s^1d^{\text{group number} - 1}$ or $d^{\text{group number}}$, the number of valence electrons is equal to the group number, and these are removed to give the maximum value of the oxidation state. More accurately, it should be stated that these elements find it energetically possible to use all their valence electrons in bonding. In valence bond theory the valence electrons would be unpaired, ready to make electron-pair bonds, with the ns and $(n - 1)$d atomic orbitals used up to a maximum of six bonds. To make more than six bonds the np atomic orbitals have to be used. The maximum number of single bonds would then be nine,

Table 5.4 Oxidation states of the transition elements; those shown in **bold red** are most stable in aqueous solution; those in *italic red* are most stable in the absence of dioxygen

Group 3	4	5	6	7	8	9	10	11	12
Sc	*Ti*	*V*	*Cr*	*Mn*	*Fe*	*Co*	*Ni*	*Cu*	*Zn*
				+7					
			+6	+6	+6				
		+5	+5	+5	+5	+5			
	+4	+4	+4	+4	+4	+4	+4		
+3	*+3*	*+3*	**+3**	+3	**+3**	+3	+3	+3	
	+2	+2	+2	**+2**	*+2*	**+2**	**+2**	**+2**	**+2**
	+1	+1	+1	+1	+1	+1	+1	+1	+1
	0	0	0	0	0	0	0		
		−1	−1	−1		−1			
			−2	−2	−2				
				−3					
			−4						
Y	*Zr*	*Nb*	*Mo*	*Tc*	*Ru*	*Rh*	*Pd*	*Ag*	*Cd*
					+8				
				+7	+7				
			+6	+6	+6	+6			
		+5	+5	+5	+5	+5			
	+4	+4	+4	+4	+4	+4	+4		
+3	+3	+3	+3	+3	**+3**	**+3**	+3	+3	
	+2	+2	+2	+2	+2	+2	**+2**	+2	**+2**
	+1	+1	+1	+1	+1	+1	+1	**+1**	+1
	0	0	0	0	0	0	0		
		−1		−1		−1			
	−2		−2		−2				
		−3							
Lu	*Hf*	*Ta*	*W*	*Re*	*Os*	*Ir*	*Pt*	*Au*	*Hg*
					+8				
				+7	+7				
			+6	+6	+6	+6	+6		
		+5	+5	+5	+5	+5	+5	+5	
	+4	+4	+4	+4	*+4*	**+4**	**+4**[a]		
+3	+3	+3	+3	+3	+3	+3	+3	**+3**[a]	
	+2	+2	+2	+2	+2	+2	+2	+2	**+2**
	+1	+1	+1	+1	+1	+1		+1	+1
	0	0	0	0	0	0	0		
		−1		−1		−1			
	−2		−2						
		−3							

[a] These oxidation states are those of the metals coordinated by ligands other than water, *e.g.* $[PtCl_4^{2-}]$ and $[AuCl_4^{-}]$.

but that would be the case if all available orbitals were used. In practice, the fluoride with highest coordination number is ReF_7, in accordance with that expected from the group number. The slightly smaller Tc atom forms TcF_6 as its fluoride of highest oxidation state, and the much smaller Mn atom forms only MnF_4. The stoichiometry is, in such cases, determined by the ease of packing ligand atoms around the central metal atom, rather than any electronic cause. Because oxides are formed in which double bonds participate (M=O), the highest oxidation states of the transition elements tend to be revealed in such compounds. The Group 7 metals all form +7 oxides of the formula $M^{VII}{}_2O_7$, in which the M atoms are tetrahedrally coordinated by only four oxygen atoms, $O_3M–O–MO_3$. The same is true of aqueous solution chemistry, where all three elements of Group 7 form MO_4^- ions.

Apart from the restrictions of coordination number, there remains the problem of why the elements of Groups 8–12 have compounds that have coordination numbers less than seven in their maximum oxidation states. The reason lies in the availability of orbitals of sufficiently low energy to make the compound formation feasible. In the Group 7 compounds the central atom has the configuration nd^7, and if two d electrons are promoted to give the configuration $(n - 1)d^5ns^1np^1$ the basis of the formation of seven electron pair bonds has been achieved. For elements of Group 8 there would have to be an extra promotion of a d electron to another p orbital to give the configuration $(n - 1)d^5ns^1np^2$; this is the basis of the +8 compounds formed by ruthenium and osmium, RuO_4 and OsO_4, although their highest oxidation state fluorides are RuF_6 and OsF_7. Continuing with the valence bond argument, the elements of Group 9 have the configuration nd^9, and to attain nine coordination there would have to be one d to s and three d to p promotions, which would mean the expenditure of too much energy for the compound formation to be feasible. These metals form +5 compounds, which are compatible with one d to s and one d to p promotions to give the configuration $(n - 1)d^7ns^1np^1$. Likewise, the Group 10 elements are d^{10}, and can achieve positive oxidation states by one d to s promotion (to produce +2 compounds), an additional d to p promotion (to produce +4 compounds), and to produce +6 compounds (observed for Pt) a second d to p promotion is necessary. Group 11 elements have the configuration $(n - 1)d^{10}ns^1$ and achieve their maximum oxidation states by one d to p promotion. The Group 12 elements simply lose their ns^2 electrons to give their +2 states. The deviations from the expected maximum oxidation states are due to the difficulties of promotion of sufficient d electrons to the higher energy s and p levels. Given that the oxidation state concept is overworked in the description of the compounds of the transition elements, the valence bond approach furnishes a better explanation of the observed maximum oxidation states and the breakdown in the regular patterns in the second and third series. Molecular orbital theory, left for other books in this series, also gives a satisfactory explanation for the observed patterns. Another factor which governs the formula of the highest oxidation state fluoride is the geometric difficulty of having sufficient room for ligand fluorine atoms around the central metal atom.

Table 5.4 shows that some of the transition elements form compounds in which the metals are in negative oxidation states. This also is an overplaying of the oxidation state concept and is best left to valence bond ideas at this stage to furnish an explanation. The negative oxidation states are formed by elements to the left side of each series, where there are fewer d electrons and where more electrons are required if the 18-electron rule is to be either obeyed or approached. For example, the compound $Na_4[Cr(CO)_4]$ contains the complex ion $[Cr(CO)_4^{4-}]$. The electron counting for the valence shell of the Cr^{-IV} (the atom is in its –4 state because the CO ligand groups are neutral molecules) is $3d^6$ from the neutral atom plus four electrons from the sodium atoms, making up the overall charge of the complex ion, plus the eight electrons donated by the four CO groups in forming four coordinate bonds, making a total of 18. Across any of the three series the necessity of adhering to the 18-electron rule by forming negative ions is a decreasing tendency, and beyond Group 9 does not occur.

The pattern of stable oxidation states in aqueous solution is different from that of the solid compounds of the transition elements. As the oxidation state of an element increases, there is a tendency, particularly in the later groups, for the compound to act as an oxidizing agent. If the compound is dissolved in water, the compound oxidizes the water to dioxygen and is consequently reduced to a lower oxidation state, one which does not have the capacity to oxidize water and which is then stable.

5.6 Oxidation States of the f-Block Elements

The observed oxidation states of the lanthanide and actinide elements are given in Table 5.5. Fifteen elements are included in both series to avoid the argument as to whether La and Ac should be included and/or whether Lu and Lr are properly members of Group 3.

5.6.1 The Lanthanides

The lanthanide elements all form +3 states, which are either the only oxidation states produced in their compounds or are the more stable state for those elements which form either +2 or +4 states. Figure 5.13 shows the relatively small ranges of the first four successive ionization energies of the lanthanide elements. The large increase after the third

Table 5.5 Oxidation states of the f-block elements; those in **bold red** are the most stable states, those in *italic red* are the most stable in aqueous solution in the absence of dioxygen

La	*Ce*	*Pr*	*Nd*	*Pm*	*Sm*	*Eu*	*Gd*	*Tb*	*Dy*	*Ho*	*Er*	*Tm*	*Yb*	*Lu*
		+4	+4						+4					
+3	**+3**	**+3**	**+3**	**+3**	**+3**	**+3**	**+3**	**+3**	**+3**	**+3**	**+3**	**+3**	**+3**	**+3**
					+2	+2							+2	

Ac	*Th*	*Pa*	*U*	*Np*	*Pu*	*Am*	*Cm*	*Bk*	*Cf*	*Es*	*Fm*	*Md*	*No*	*Lr*
				+7	+7									
			+6	+6	**+6**	+6								
		+5	+5	**+5**	+5	+5								
	+4	+4	*+4*	+4	+4	+4	+4	+4	+4					
+3	+3	+3	+3	*+3*	*+3*	**+3**	**+3**	**+3**	**+3**	**+3**	**+3**	**+3**	**+3**	**+3**
						+2			+2	+2	+2	+2	*+2*	

ionizations is responsible for the normal oxidation state maximum. The elements are all very electropositive, with electronegativity coefficients of 1.1 (*cf.* Na, 1.0), and form ionic compounds. The oxidation states other than +3 are generally less stable than the +3 state and only Ce^{IV}, Eu^{II} and Yb^{II} are reasonably stable. Of the other "exceptions", Pr^{IV} and Tb^{IV} are very strong oxidizing agents and Sm^{II} is a very strong reducing agent, so that they do not have an independent existence in aqueous solution.

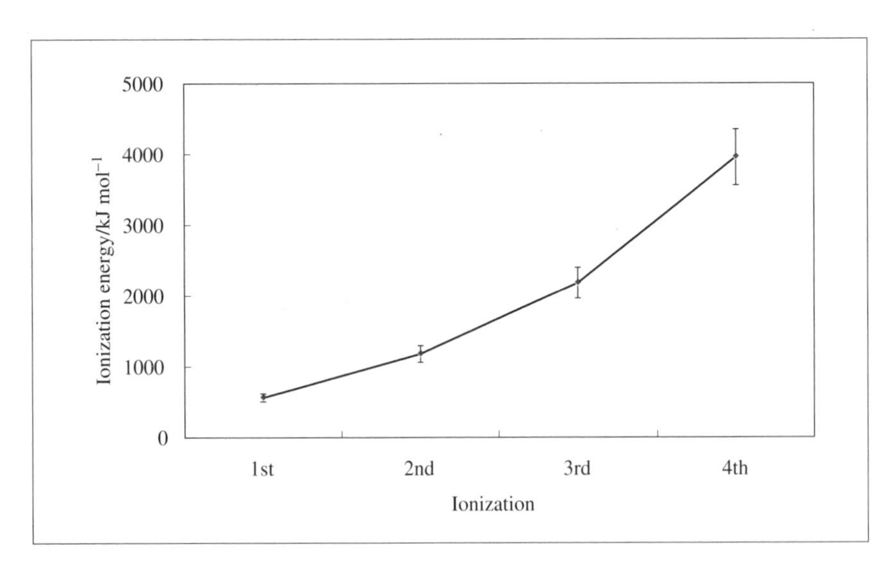

Figure 5.13 A plot showing the ranges of the first four successive ionization energies of the lanthanide elements

5.6.2 The Actinides

In comparison with the lanthanides, the actinides exhibit a greater range of oxidation states, as indicated in Table 5.5. The early members of the series from Ac to Cm show a greater similarity to the left-hand side of a series of transition elements, but after Cm the later members of the series are much like the corresponding lanthanide elements. The multiplicity of oxidation states of the early actinides is due to the lower ionization energies of the 5f electrons. In the second half of the series the 5f ionization energies are too large for any higher oxidation states to be stable. Those elements which show a range of oxidation states are much more easily separated than are the lanthanide elements, oxidation and reduction reactions facilitating the preferential precipitation of one or other of the elements from aqueous solutions.

Summary of Key Points

1. The concepts of valency and oxidation state were defined and exemplified.

2. The periodicity of valence of the s- and p-block elements was described and rationalized in terms of the octet rule. Exceptions from the octet rule were discussed. These include compounds exhibiting hypervalence as a result of the expansion of the valence shell of the central atom, and compounds in which the inert pair effect is apparent, the valency of the central element being two units lower than expected for the group valency.

3. The periodicities of the oxidation states of the d-block and f-block elements were described.

Problems

5.1. Non-metallic elements which exist as: (i) diatomic molecules, (ii) small polyatomic molecules, (iii) polymeric chains, (iv) polymeric sheets and (v) infinite three-dimensional arrays should have very different melting points. Explain why this should be so and give examples to support your argument.

5.2. What are the oxidation states of the atoms in the ammonia molecule?

5.3. What are the oxidation states of the atoms in the nitrogen trifluoride molecule? Comment on the valencies of nitrogen in ammonia and NF_3.

5.4. Interpret the melting points (in K) of the elements given below in terms of their structures.

B	*C (diamond)*	*N*	*Al*	*Si*	*P (white)*
2573	3820	63	934	1683	317

5.5. What are the oxidation states of Pt and Au in the complex ions $[PtCl_4^{2-}]$ and $[AuCl_4^-]$?

Further Reading

S. E. Dann, *Reactions and Characterization of Solids*, Royal Society of Chemistry, Cambridge, 2000. A companion volume in the tutorial text series. This book contains a more extensive treatment of metallic bonding than that given in this text.

J. Barrett, *Structure and Bonding*, Royal Society of Chemistry, Cambridge, 2001. A companion volume in the tutorial text series.

N. C. Norman, *Periodicity and the s- and p-Block Elements*, Oxford Science Publications, Oxford, 1997.

D. M. P. Mingos, *Essential Trends in Inorganic Chemistry*, Oxford University Press, Oxford, 1998.

C. J. Jones, *d- and f-Block Chemistry*, Royal Society of Chemistry, Cambridge, 2001. A companion volume in the tutorial text series. This has an account of Russell–Saunders coupling and a more detailed survey of the oxidation states of the d- and f-block elements.

6

Periodicity III: Standard Enthalpies of Atomization of the Elements

The subject of this chapter is the periodicity of the standard enthalpies of atomization, $\Delta_a H^{\ominus}$, of the elements, and their correlation with valency and the number of valency electrons.

The standard enthalpy of atomization is a good measure of the stability of an element in the form it exists under standard conditions. The values range from zero for the Group 18 elements, which are already monatomic gases in their elementary state, to tungsten with a value of +849 kJ mol^{-1}, indicating that its atoms are the most tightly bound of all the elements The normal thermodynamic convention is to base all calculations on a reference level, which is that the standard enthalpies of the elements in their standard states are *zero*. This hides the reality of the elements having very different stabilities with respect to their gaseous atoms.

The valency electrons, experiencing their various effective nuclear charges and interactions with each other, determine the structure of the individual elements as the total energy is minimized. Some structures are described in the text, but the detailed periodicity of elementary forms is left to more comprehensive texts.

The oxidation state concept is not needed in this chapter because all elements in their elemental states are in their zero oxidation state by convention. The standard values are for the formation of one mole of the gaseous monatomic element at 298 K and 101325 N m^{-2} pressure.

Aims

By the end of this chapter you should understand:

- What is meant by the standard enthalpy of atomization of an element
- The variations of enthalpy of atomization along the periods and down the groups of the Periodic Table
- The connections between valency and the value of the standard enthalpy of atomization of an element

6.1 Periodicity of the Standard Enthalpies of Atomization of the Elements

Figure 6.1 shows plots of the standard enthalpies of atomization for Periods 2–6. The increases in the values from Group 1 to Group 2 are followed by a progression through a maximum and a decrease to zero from Group 13 to Group 18. The intervening three transition series in Periods 4–6 show a general pattern, the values rising from Group 3 to maxima around Groups 5 and 6 and then decreasing to the Group 12 values. In the first transition series there is a dip in the curve with a minimum at manganese.

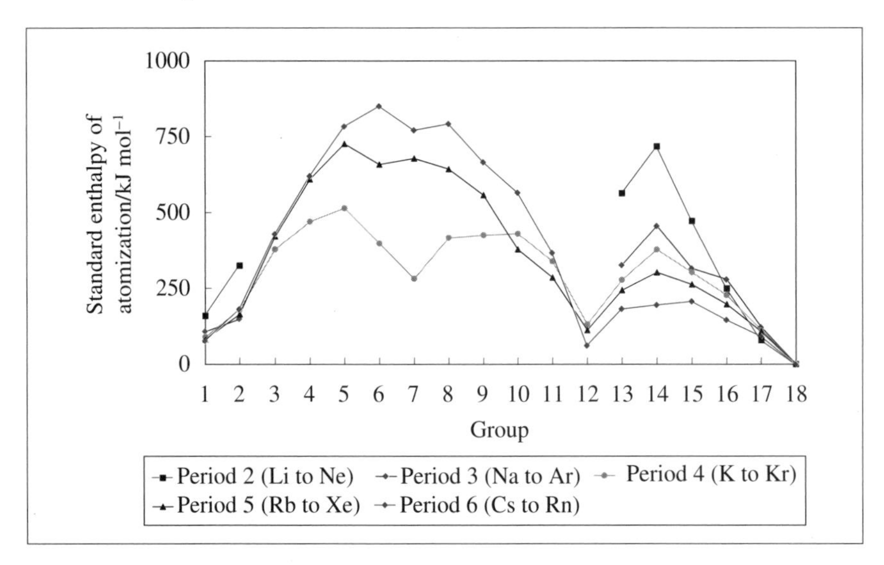

Figure 6.1 The standard enthalpies of atomization for Periods 2 to 6

The values may be understood in terms of the cohesive forces operating in the various cases. The basis of the theory is that which is presented in Section 4.3 in the discussion of the atomic volume curve. The forces range from metallic bonding in Groups 1 to 12/13, followed by covalent bonding in the p-block non-metallic elements, to van der Waals forces (London interatomic forces) which operate alone in the Group 18 elements.

The overall process of atomization is made up from various contributions:

(i) The melting of solid elements brought about by the loosening of metallic bonds and the breaking of van der Waals forces between molecules.

(ii) The conversion of the liquid elements to the gas phase, accompanied by the atomization of any molecular species brought about by the final breaking of metallic bonds and residual van der Waals forces, and by the dissociation of molecules into atoms.

The detailed discussions of the s- and p-block and the d-block elements are separated in the following sections, and finally the trends in the lanthanide and actinide elements are described and discussed.

6.1.1 Variations of $\Delta_a H^\ominus$ in the s- and p-Block Elements

In the first period (H and He), the comparison between the diatomic dihydrogen and monatomic helium shows the large difference between the strength of the single covalent bond in dihydrogen and the London interatomic forces in helium. Helium is a gas under standard conditions and its standard enthalpy of atomization is therefore zero. Dihydrogen is also a gas under standard conditions, but the single bond requires 436 kJ (mol H_2)$^{-1}$ to cause its dissociation into two hydrogen atoms. The standard enthalpy of atomization of hydrogen (as are the values for the other elements) is quoted in terms of the enthalpy required to produce *one mole of atoms, i.e.* $436 \div 2 = 218$ kJ (mol H)$^{-1}$.

Figure 6.2 brings together the values of the enthalpies of atomization for the s- and p-block elements. In Period 2 there is a transition from metallic lithium and beryllium to the p-block elements B, C, N, O and F, which are covalently bonded with the molecular units, if any, held together by van der Waals forces.

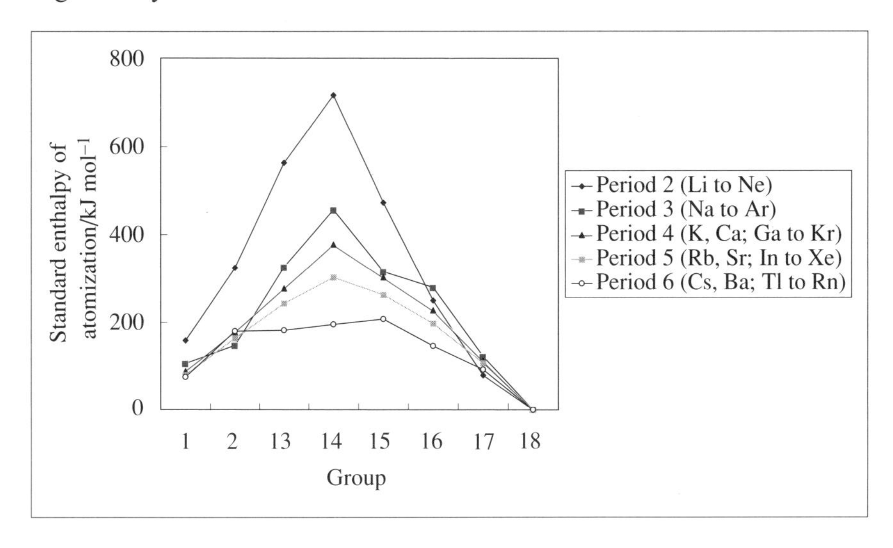

Figure 6.2 The standard enthalpies of atomization for the s- and p-block elements

Lithium is a soft metal with a melting point of 453 K. Its crystal structure is body-centered cubic, with each lithium atom coordinated by eight nearest neighbours. There is one valency electron, $2s^1$ from each atom, so the bonding is essentially weak. The 2s orbitals overlap their neighbours and form an infinite band of molecular orbitals, half of which (the lower bonding half) are occupied by the available valency electrons.

Beryllium has a hexagonal close-packed structure in which the coordination number for each atom is 12. As in Li, the 2s orbitals overlap to give an infinite band which in the $2s^2$ beryllium atoms would be full. This would not lead to cohesion, and bonding is brought about by the

Worked Problem 6.1

Q Hydrogen and lithium both have a single valence electron. Why does lithium not form a Li_2 molecule in its standard state?

A Lithium vapour does contain Li_2 molecules, but the 2s valence electron in the Li atom is efficiently shielded from the effect of the nuclear charge by the inner $1s^2$ pair of electrons, so the covalent bond in Li_2 is long (267 pm) and weak (107 kJ mol^{-1}). With Li a metallic lattice is formed in which the effects of the assembly of the valence electrons combine to give better stability than in the diatomic molecule. The distance between nearest neighbours in the metal is 304 pm, but there are eight such neighbours, and this leads to a stabilization of 161 kJ mol^{-1}.

2p band overlapping with the 2s band so that the bonding parts of both bands are occupied, so making Be more stable than Li.

In the Group 13–17 elements there are more electrons available for bonding, and covalency becomes preferable to metallic bonding, but in varied ways. Boron has four valency orbitals, 2s and the three 2p, but only three valency electrons, $2s^2 2p^1$. This is described as **electron deficiency**, and applies not only to boron but also to most elements which form the metallic state. In boron, the ionization energies are relatively high and the 2p–2s energy gap precludes the overlapping of possible metallic-type bands, so boron forms complex structures which maximize the possibilities of covalency with insufficient electron pairs. The two allotropes of boron, the α- and β-forms, are both based upon an icosahedral B_{12} unit. This is shown in Figure 6.3.

The complexities of structures such as those of elementary boron are better appreciated after studies of symmetry theory and cluster chemistry, and so are not elaborated in this text.

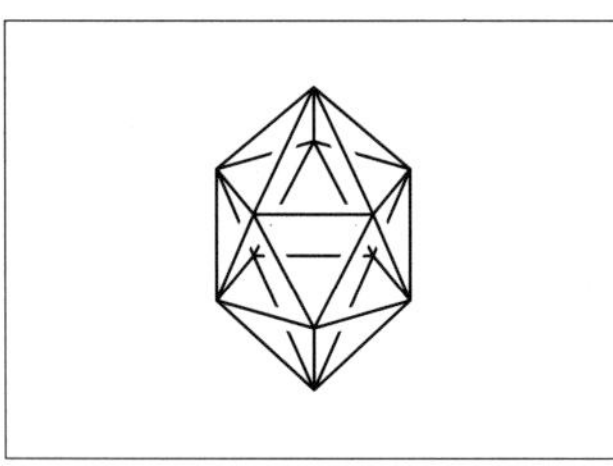

Figure 6.3 The icosahedral B_{12} unit of boron structures

The α-form consists of a cubic close-packed arrangement of icosahedra which are linked by bonds in which three boron atoms and two electrons participate. If three boron atoms supply one orbital each, the orbitals form three molecular orbitals, one of which is bonding and can accommodate the two available electrons. In spite of the relatively weak bonding between the icosahedral units, the three bonding electrons per atom bestow sufficient cohesion on the various allotropes of boron to make the enthalpy of atomization considerably greater than that of beryllium.

The carbon atom has four valence electrons and four valency orbitals. To that extent it is **electron precise**, and can form four bonds per atom. It does this in its various allotropes, graphite being its standard state. The structure of graphite is shown in Figure 6.4 and consists of hexagonally bonded sheets of carbon atoms that are 335 pm apart. The carbon–carbon distances in the sheets are all 141.5 pm between nearest neighbours, consistent with a bond intermediate between single (C–C) and double (C=C) bond length, and almost identical with the C–C bond distance in the benzene molecule. The sheets are built up in an ABAB... formation, although there are forms of graphite with ABCABC... formations and more complex, less orderly, arrangements.

In organic compounds generally, carbon–carbon single bonds have a length of 154 pm and double bonds are 134 pm long.

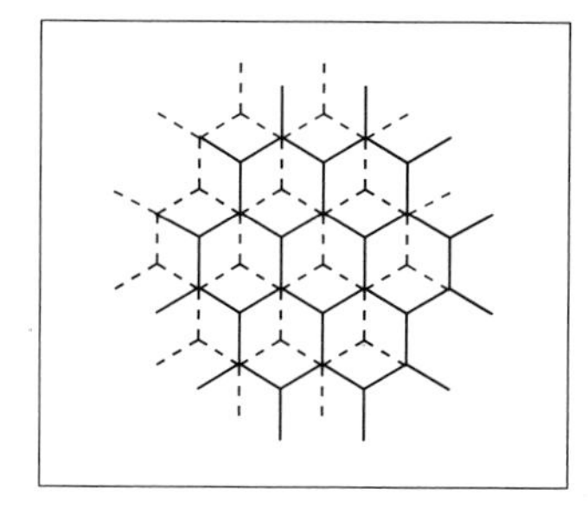

Figure 6.4 The structure of graphite showing the relative positions of two layers

There is a weak electronic interaction between the π-systems of adjacent sheets and that, with van der Waals forces operating, holds the sheets together. The delocalization of the π electrons within each sheet provides the stability of the structure, which is 190 J mol^{-1} more stable than the other principal allotrope, diamond. In diamond, each carbon atom exerts its valency of four in a tetrahedral manner, and the carbon–carbon distance is 154 pm, *i.e.* that expected for a single covalent bond. The enthalpy of atomization of carbon is significantly larger than that of boron, reflecting the effect of the extra valence electron.

In elementary nitrogen, two of the valence electrons are paired up in the 2s atomic orbital and only three electrons are available for bonding. There is no way of unpairing the 2s electrons, as the next lowest energy level is the 3s which is too high to be useful. The most stable form of bonding in elementary nitrogen is found in the triple covalency of the dinitrogen molecule. The N≡N bond has a dissociation energy of 944 kJ mol^{-1}, and is one of the strongest multiple covalent bonds known. Because there are only three electrons per atom responsible for the bonding, the enthalpy of atomization of nitrogen shows a decrease compared to that of carbon.

Worked Problem 6.2

Q Why is there such a dramatic difference in elementary form between carbon and nitrogen, moving from infinite array (diamond, graphite) to diatomic molecule, N_2?

A The nitrogen atom has a higher effective nuclear charge than the carbon atom, making covalent bonding more effective, but it also has a pair of electrons which do not participate in the bonding. When this occurs there is a significant weakening of the atom–atom single bond enthalpy term, so rather than forming an infinite array, as does boron with three valence electrons, nitrogen forms the triple bonded diatomic molecule. The C–C, N–N, O–O and F–F single **bond enthalpy terms** are 348, 163, 146 and 158 kJ mol^{-1}. The latter three bonds are weakened because of the presence of lone pairs of electrons on the two participating atoms, which repel each other. The effect is absent in the single C–C bond.

A bond enthalpy term is the average bond enthalpy for a particular bond, and applies to all the molecules in which the bond is found.

The decrease continues with oxygen and fluorine, which form diatomic molecules O_2 and F_2 in their standard states. The 2p electrons of both oxygen atoms are associated with the formation of a bond with a bond order of 2 and in F_2 the bond is a single one, consistent with the sole unpaired electron in the free F atom. The single bond in F_2 is particularly weak because of repulsions between the three lone pairs of electrons on both of the two atoms.

Worked Problem 6.3

Q The bond dissociation enthalpies of the halogen molecules F_2, Cl_2, Br_2 and I_2 are 158, 243, 193 and 151 kJ mol^{-1}, respectively. Is it possible that the bonds in these molecules, apart from that in F_2, have bond orders greater than one?

A Yes, a possible explanation for the value for difluorine being anomalously low is that the other halogens have available d orbitals and that a p to d promotion would allow triple bond formation.

With neon the end of the 2nd period is reached and that element, with its $2s^22p^6$ octet, does not form chemical bonds. The element exists in its standard state as a gas. Its enthalpy of atomization is zero.

The elements of the 3rd period show a similar pattern to those of the 2nd period, their enthalpies of atomization going through a maximum at silicon and decreasing to zero with argon. There are some differences between the two sets of elements; the enthalpies of atomization show a general decrease down any group and there are some variations in the structures of the elements.

Sodium and magnesium are similar to Li and Be in the 2nd period, and have one and two valence electrons, respectively, which participate in metallic bonding.

The first large difference between the two periods is with aluminium, which is metallic with a cubic closest-packed structure. In going down any group there is a decrease in the effective nuclear charge, which ensures an increase in the tendency towards metallic bonding rather than covalent bonding, the latter being most efficient when the effective nuclear charge is high. The standard state of silicon is its diamond-like structure, with four single Si–Si covalent bonds per atom, which are weaker than those between carbon atoms. The element has a blue-grey metallic lustre and is a semiconductor, again showing the transition from non-metallic to metallic character going down a group, silicon being at an intermediate stage. Elementary phosphorus has at least five allotropic forms, and the one which is taken to be the standard state is white phosphorus, which consists of P_4 tetrahedra. The red and black forms of the element are infinite arrays, which are thermodynamically more stable than the white form, but are less well characterized and so neither is considered to be the standard state of the element. The tendency towards multiple bonding, so evident in dinitrogen, is not present in P, which forms P_4 tetrahedra with P–P single bonds.

Worked Problem 6.4

Q Boron and phosphorus both possess three valence electrons, but they have very different structures in their elementary forms. Why is this?

A The main reason for the difference is that the boron atom is electron deficient, and forms structures based upon icosahedral B_{12} clusters which allow each boron atom to have a share of more electrons. Multiple bonding in phosphorus is weak, and the element exerts its valency of three in the P_4 tetrahedral clusters which form the standard state.

With sulfur the standard state is the α-rhombic form, which consists of S_8 molecules held together by van der Waals forces. The multiple bonding found in O_2 is rejected in favour of the ring formation using S–S single covalent bonds. There are other allotropic forms of the element, all depending upon ring formation, except plastic sulfur which has entangled long chains of variable length. Chlorine forms diatomic molecules similar to F_2, but the bond strength of Cl_2 is greater than that of F_2. The weakness of the F–F single bond is thought to be due to the repulsions between the non-bonding electrons in the valence shells of the two atoms, although there is a possibility that the other halogen molecules might possess some multiple bonding, via their d orbitals. The 3rd period ends with the Group 18 gas argon, with a zero value for its enthalpy of atomization.

Going down the groups there is a general decrease in the values of the enthalpy of atomization of the elements, but there are exceptions in Groups 2, 16 and 17. That in Group 17 is referred to above. In Group 2 the values alternate: Be (324 kJ mol^{-1}), Mg (147.7 kJ mol^{-1}), Ca (178.2 kJ mol^{-1}), Sr (164.4 kJ mol^{-1}) and Ba (180 kJ mol^{-1}). In general, bonding becomes weaker with larger participating atoms, and the differences in metallic bonding in the Group 2 metals is probably because of the differing relative energies of the s and p bands and d band participation in the elements Ca, Sr and Ba.

The Group 16 anomaly, with the oxygen value being lower than that of sulfur, is due to the high strength of the S–S single bond (the S–S single bond enthalpy term is 264 kJ mol^{-1}) and the standard state of the sulfur being the S_8 molecule in its solid state. Compared to the atomization of dioxygen, which requires the dissociation of the double bond (dissociation enthalpy = 496 kJ mol^{-1}), that of octasulfur requires the melting and evaporation of the molecule in addition to the dissociation of the S–S single bonds.

Worked Problem 6.5

Q The enthalpy of atomization of sulfur is 277 kJ mol^{-1}. Calculate the enthalpy change during the atomization process which does *not* involve the breaking of covalent bonds. With what other processes is this extra enthalpy linked?

A The enthalpy change for the atomization of the octasulfur molecule is $8 \times 277 = 2216$ kJ mol^{-1}. The enthalpy required to break the eight S–S covalent bonds in the molecule is $8 \times 264 = 2112$ kJ mol^{-1}. The difference of $2216 - 2112 = 104$ kJ mol^{-1} is the enthalpy change required to overcome the intermolecular forces between the molecules in the solid sulfur as it goes through the liquid and gaseous phases as S_8 before undergoing dissociation into atoms. This represents $(104 \times 100) \div 2216 = 4.7\%$ of the total atomization enthalpy. [Dioxygen is already a gas at 298 K and has no contributions from non-valence forces to its enthalpy of atomization.]

Although there is no change of order in the values of the enthalpies of atomization of the Group 15 elements, the value for phosphorus seems unusually low. This is probably because of the inherent weakness of the bonding in the P_4 tetrahedra, associated with the very unusual 60° bond angles.

Relativistic effects are shown by the p-block elements of the 6th period (in particular by lead); their enthalpies of atomization are considerably lower than the equivalent elements of the 5th period, and this is attributable to the relative unavailability of the $6s^2$ pair of electrons for the metallic bonding.

Worked Problem 6.6

Q Explain why the value of the enthalpy of atomization of barium seems not to show a major relativistic effect.

A Relativistic effects are dependent upon atomic number, and become noticeable in the lanthanide series and the elements beyond that series. They affect the enthalpies of atomization of the p-block elements of the 6th period.

6.1.2 Variations of $\Delta_a H^\ominus$ in the d-Block Elements

Figure 6.5 shows the variations in the enthalpies of atomization of the three series of transition elements. There is a general pattern going through a maximum, with the early and late elements in each series having relatively low values of $\Delta_a H^\ominus$, with intermediate minima at manganese, molybdenum and rhenium. The values for the Group 3 elements are greater than the corresponding values for the Group 2 elements, which have a maximum of two valency electrons that can occupy the overlapping s, p and d metallic conduction bands.

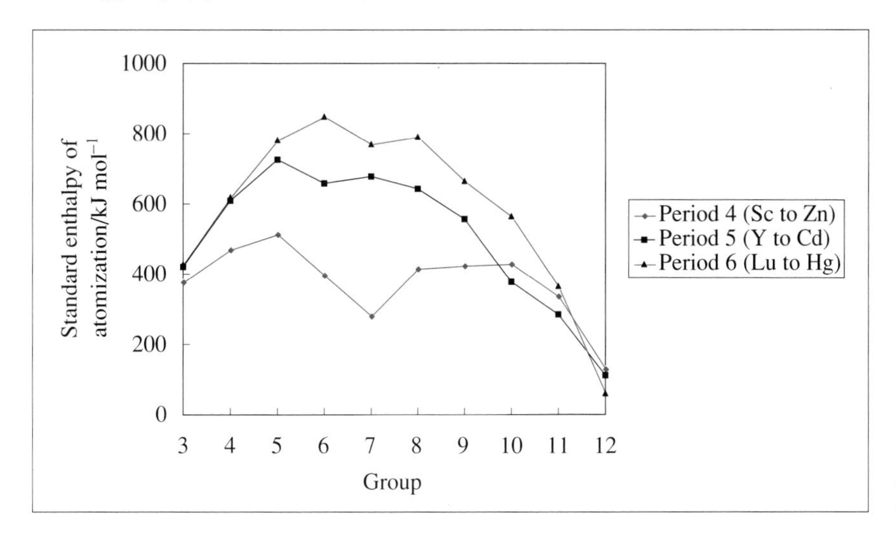

Figure 6.5 The enthalpies of atomization of the transition elements

Worked Problem 6.7

Q How may a Group 2 element with an electronic configuration s^2 use the electrons to achieve metallic bonding?

A If the s and p bands of orbitals overlap appropriately, it is possible for both electrons to occupy bonding regions of the two bands and to allow metallic bonding to be maximized. In the heavier elements it is possible that the d bands participate in stabilization.

The variations of the enthalpies of atomization of the three series of transition elements are not easily interpreted except in a non-quantitative way. The Group 3 elements have enthalpies of atomization that are significantly larger than the corresponding Group 2 elements. A reasonable conclusion is that their three valency electrons contribute fully to the metallic binding, which could be due to the use of the d and s bands. The filling of bands of orbitals in metallic structures follows the *aufbau* principle, with the most bonding orbitals filled first, and then there is a filling of orbitals which have progressively less bonding and more anti-bonding character, and last of all to be filled are the completely anti-bonding orbitals. Because the orbitals in any band are very close together, some single filling of orbitals takes place when there is a partial filling of the band. Interelectronic repulsion causes these exceptions to the aufbau principle. In the transition series, there are possibilities of filling the bands formed from the $(n - 1)$d, ns and np orbitals, and the order of filling determines the pattern of bonding across any one period.

The three plots shown in Figure 6.5 differ from each other in detail, although the overall trend is for the values to go through a maximum as the bonding parts of the metallic bands are filled, after which there is a decrease to the Group 12 element. There are three main reasons for the differences between the three plots. One is referred to above, which is that successive additions of electrons to bands do not produce identical effects on the metal–metal bonding because different parts of the bands are being filled varying from completely bonding through a region which is partially bonding, partially anti-bonding to the upper energy regions which are completely anti-bonding. The second reason for differences is the uncertainty about the participation of the p orbitals of the atoms in metallic bands. The third reason for the differences is that the gaseous atoms produced in the atomization process have varying stabilities, which are affected by the numbers of parallel-spin electrons associated with differing amounts of exchange energy. The intermediate minimum in the plot for the first series of transition elements at manganese, and the apparently low value for chromium, are associated with the energy of the gaseous atoms which are stabilized by the exchange energy of the d^5 configuration in both cases. That the value for manganese is even lower than that for chromium is probably due to the extra electron in the metal occupying an anti-bonding region of a band, s or d. After manganese in the first transition series, the enthalpy of atomization rises for Fe, remains at the same level for Co and Ni and becomes progressively smaller for Cu and Zn. The variation in atomization enthalpy in the series Fe, Co and Ni cannot be explained without some use of the p band of orbitals, which allows electrons to enter the p bonding levels. Otherwise the electrons would have to occupy the d band anti-bonding levels. In the case of zinc the bonding is due to the partial filling of the s and p bands.

In the second and third series of transition elements there are similar variations with intermediate minima at Mo and Re, respectively, but the values in general are higher than those for the first series of transition elements. This is because the 4d and 5d orbitals overlap more efficiently, and more overlap equates with stronger bonding. The 3d orbitals are relatively compact and produce a weaker bonding effect.

6.1.3 Variations of $\Delta_a H^\Theta$ in the f-Block Elements: Lanthanides and Actinides

The Lanthanides

Figure 6.6 shows the variations of $\Delta_a H^\Theta$ for the lanthanide elements, La to Lu, and the diagram also includes a plot of the third ionization energies of the elements. The first and second ionizations of the lanthanides are those in which the two 6s electrons are removed. The range of values of $\Delta_a H^\Theta$, when compared to those of the transition elements, are consistent with between two and three electrons per atom contributing to the metallic bonding. The two plots in Figure 6.6 are almost perfect mirror images. That of the 3rd ionization energies indicates the relative difficulty of removing the third electron from the lanthanide elements, which is also related inversely to the extent to which these third electrons participate in the metallic bonding. The three valence electrons of the lanthanum atom participate in the metallic bonding of the element, but across the series from La to Eu there is a progressive tendency towards only two electrons taking part in the bonding. In gadolinium, possibly because of the enforced pairing of the 4f electrons, there is a return to the three-electron bonding. After Gd, there is again a progressive tendency towards two-electron bonding, followed at the end of the series by lutetium with its $6s^2 5d^1$ configuration which allows the return to three-electron bonding. The elements for which there seems little doubt that three-electron bonding occurs are the ones with 5d electrons in their free-atom configurations: La ($6s^2 5d^1$), Gd ($6s^2 5d^1 4f^7$) and Lu ($6s^2 5d^1 4f^{14}$). The intermediate elements which do not possess a 5d electron in their free-atom configurations are less tightly bound in their metallic states, but seem to have the partial participation of a third electron in the s/d bands which provide the cohesion. This third electron would be produced by promotion from the 4f orbitals. It is clear that, apart from the promotion of possibly one 4f electron to the d band in some cases, the 4f electrons play no part in the bonding of the lanthanide metals.

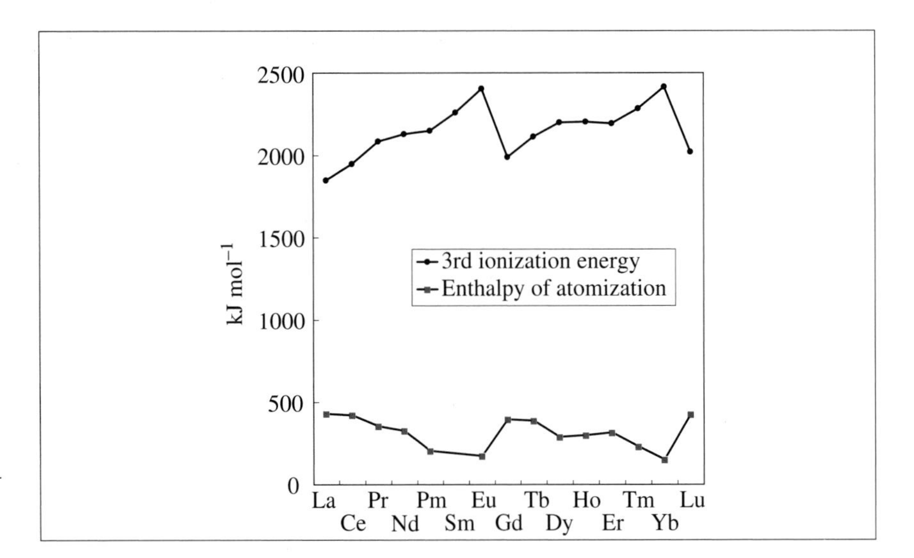

Figure 6.6 The enthalpies of atomization of the lanthanide elements, La to Lu, and their third ionization energies

Worked Problem 6.8

Q Explain the discontinuities in the plot of the third ionization energies of the lanthanides shown in Figure 6.6.

A The third ionization energy of europium is the energy required to cause the change in electronic configuration of [Xe]$4f^7$ to [Xe]$4f^6$. The process is accompanied by the loss of exchange energy of $6K$, which inflates the value. The third ionization of gadolinium is from the configuration [Xe]$4f^8$ to [Xe]$4f^7$, in which there is no change in exchange energy, but the process is made easier by the electron which is removed coming from a filled 4f orbital.

The Actinides

The actinides differ from the lanthanides in having electronic configurations in which d orbitals feature more in some of their valence shells. Actinium has the same outer configuration, s^2d^1, as does lanthanum, but whereas the lanthanides after lanthanum dispense with the use of the 5d levels (apart from Gd which is $6s^25d^14f^7$), the next actinides do have one or two 6d electrons in their free-atom configurations: Th ($7s^26d^2$), Pa ($7s^26d^15f^2$), U ($7s^26d^15f^3$) and Np ($7s^26d^15f^4$). Plutonium and the other later elements of the series have configurations which are lanthanide-like, with only curium having a single 6d electron (*cf.* Gd). These free-atom electron configurations are relevant to the understanding of the $\Delta_aH^\ominus$ values of their metallic states. The data that are available are plotted in

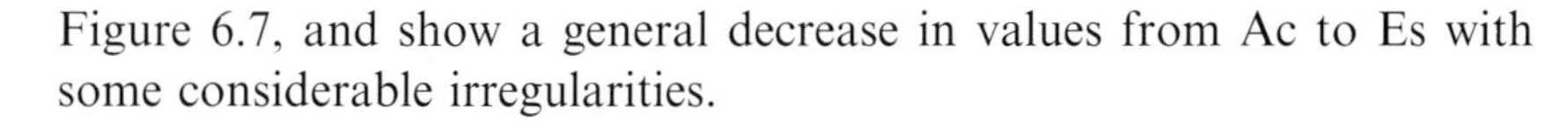

Figure 6.7, and show a general decrease in values from Ac to Es with some considerable irregularities.

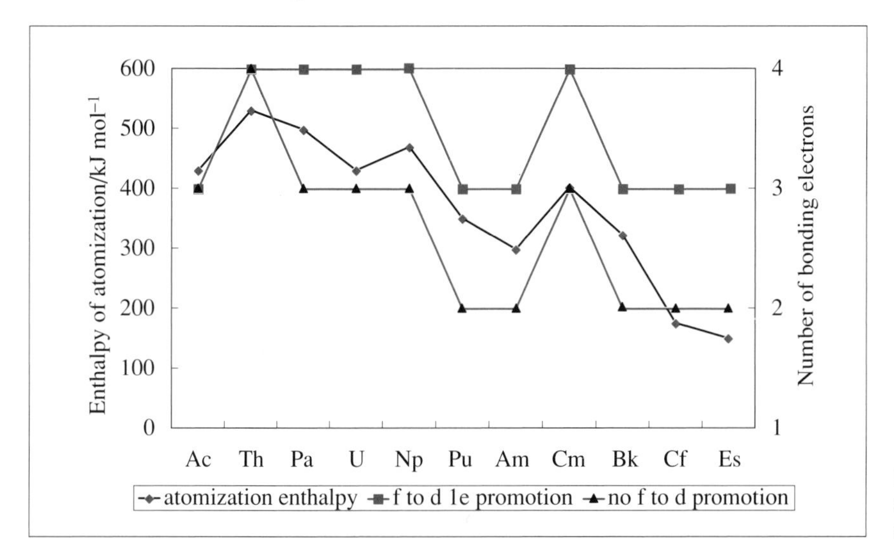

Figure 6.7 The enthalpies of atomization of the actinide elements, Ac to Es

The value for Ac is consistent with a three-electron bonding contribution to the s/d metallic bands, and in thorium it would seem that there are four electrons per atom contributing to the binding of the metal. There is then a somewhat irregular decrease to einsteinium, which has a two-electron contribution to its metallic bonding, similar to that of europium. The deviations are significant enough to suggest that some f to d promotion is occurring in Pa, U and especially with Np. The free-atom 6d electron in curium participates in the metallic bonding of the element, but after Cm the $\Delta_a H^\ominus$ values subside to the two-electron contributions to the s/d bands expected from their $7s^2$ free-atom configurations with no contributions from the 5f electrons. At the end of the series lawrencium, with its full 5f shell, a single 6d electron and the $7s^2$ pair, would be expected to have an enthalpy of atomization consistent with a three-electron contribution to the s/d bands.

The data in Figure 6.7 are shown with two theoretical plots of possible numbers of electrons contributing to the metallic bonding. The upper red line is the plot of the numbers of bonding electrons, assuming that in Pa, U, Np and Cm there are f to d promotions which allow four electrons to occupy the s/d bands which cause the bonding, and that in Pu, Am, Bk, Cf and Es there are similar promotions that increase the number of bonding electrons from two to three. The lower red line is the plot of the number of electrons in the s/d bands assuming that no promotions occur. The two theoretical plots enclose the real data almost completely and suggest that the promotion theory is acceptable. A recent report[1] has given a detailed explanation of the extraordinary expansion of 25% in volume that occurs when the low-temperature form of plutonium metal (the α-phase stable below 400 K) is converted to the high-temperature δ-phase (stable around 600 K). It is suggested that the bonding in the δ-phase is that produced by two bonding electrons per atom, whereas that of the denser α-phase is due to three bonding electrons per atom.

Summary of Key Points

1. The periodicity of the enthalpies of atomization of the elements was described and discussed in terms of the s- and p-blocks, the three transition series, and the lanthanide and actinide elements.

2. The periodicities were rationalized in terms of either metallic bonding or covalent bonding, and related to the electronic configurations of the free atoms of the elements.

Problems

6.1. The standard enthalpy of atomization of white phosphorus is 316.5 kJ (mol P)$^{-1}$. The enthalpy of sublimation (*i.e.* for the conversion of the molecular solid into the molecular gas) of the element is 58.9 kJ (mol P_4)$^{-1}$ Calculate the P–P bond enthalpy. What processes are represented by the enthalpy of sublimation? Calculate the proportion of the total enthalpy of atomization which is used in the sublimation of the solid P.

6.2. The standard enthalpy of atomization of iodine is 107 kJ (mol I)$^{-1}$. The I–I dissociation enthalpy is 151 kJ mol^{-1}. Calculate the contribution to the enthalpy of atomization caused by the dissociation of the covalent bond, and comment on the causes of the remainder.

6.3. The elementary form of diiodine exists as almost black crystals with a metallic sheen. Is iodine showing metallic character?

6.4. Although melting points and boiling points of the elements show periodicities that are similar to the variations of enthalpies of atomization, they are not dealt with in this book because of shortage of space. It would be a good exercise for the reader to enquire into such periodicities. For this problem, the melting points and enthalpies of atomization of the first transition series are given below. Is there a good correlation between the two properties for the elements concerned? What conclusion is possible from the correlation? Would a better correlation be expected with the boiling points of the elements?

Element	*m.p.*/°C	$\Delta_a H^{\ominus}$/kJ mol^{-1}	*Element*	*m.p.*/°C	$\Delta_a H^{\ominus}$/kJ mol^{-1}
Sc	1540	340	Fe	1535	418
Ti	1675	469	Co	1492	427
V	1900	515	Ni	1453	431
Cr	1890	398	Cu	1083	339
Mn	1240	279	Zn	420	130

6.5. Explain the discontinuity in the plot of the third ionization energies of the lanthanides at lutetium in Figure 6.6.

Reference

1. S. Y. Savrasov, G. Lotliar and E. Abrahams, *Correlated electrons in δ-plutonium within a dynamical mean-field picture*, in *Nature*, 2001, **410**, 793. A report about the low density of the δ-phase of plutonium metal.

Further Reading

D. M. P. Mingos, *Essential Trends in Inorganic Chemistry*, Oxford University Press, Oxford, 1998.

P. W. Atkins, *Physical Chemistry*, 6th edn., Oxford University Press, Oxford, 1998.

J. Emsley, *The Elements*, 3rd edn., Oxford University Press, Oxford, 2000.

7

Periodicity IV: Fluorides and Oxides

Hydrogen combines with many elements, and the periodicity of the hydrides is of interest. Hydrides are not formed by a number of the transition elements, and for that reason the topic of hydride periodicity is not included in this text.

The two most electronegative elements, fluorine from Group 17 and oxygen from Group 16, form binary compounds (*i.e.* there are no other elements present) with almost all of the elements. The simpler compounds, *e.g.* usually those with only one atom of the other element, are used in this study of the chemical periodicity of the elements. In addition, most of the oxides of the elements react with water to give either alkaline or acidic solutions, and this is an almost universal property that shows periodicity across the periods and down the groups of the periodic table.

Aims

By the end of this chapter you should understand:

- The periodicity of the fluorides of the elements
- The periodicity of the oxides of the elements
- The formation of acid or alkaline solutions when oxides dissolve in water

7.1 The Fluorides and Oxides of the Elements

Fluorine has the capacity to oxidize most elements to their maximum oxidation state; in the cases where this does not occur, it is because of the steric difficulties of arranging sufficient numbers of fluorine atoms around the central atom. For instance, the highest oxidation state (VII) of osmium produced by reaction with fluorine is in OsF_7, but the compound is unstable at temperatures above 173 K and decomposes to give the hexafluoride. Dioxygen, a potentially less good oxidizing agent than fluorine, oxidizes osmium to the tetraoxide, OsO_4, since the steric difficulties are absent and four oxygen atoms can approach the central osmium atom to make efficient *double* covalent bonds in the Os^{VIII} compound.

The other halogens are less good at oxidizing elements, so there are smaller ranges of chlorides, bromides and iodides than fluorides.

Worked Problem 7.1

Q The shapes of the OsF_7 and OsO_4 molecules are pentagonal bipyramidal and tetrahedral, respectively. What are the angles formed by the Os–F and Os–O bonds in the two compounds?

A In OsF_7, the two Os–F bonds in apical positions are at 180° to each other, diametrically opposed. The apical bonds are at right angles to the bonds in the pentagonal plane, and the adjacent bonds in the plane form angles of 72°, the latter being a destabilizing factor in the structure. Such a small bond angle tends to make the adjacent F atoms too close to each other. In OsO_4, the bond angles are the regular tetrahedral angles of 109°28′, so any steric interaction between the ligand oxygen atoms is minimized.

7.2 Fluorides of the Elements

7.2.1 Hydrogen Fluoride and the Other Hydrogen Halides

Hydrogen forms the covalent halides HF, HCl, HBr and HI which, as isolated molecules, have polarities in accordance with the differences in the electronegativity coefficients of the halogens and that of hydrogen. In their standard states the four hydrogen halides are gases. In their solid states they form zigzag chains due to hydrogen bonding between adjacent molecules, and this persists in the liquid state. Hydrogen bonding is at a maximum in HF, as indicated by its exceptionally high boiling point of 19.7 °C, whereas those of the other hydrogen halides are considerably lower: HCl, –85.1°C; HBr, –66.8 °C; and HI, –35.4 °C. The gases dissolve in water to give acidic solutions, as the covalent bonds break so that the halogen atom retains the valence electron of the hydrogen atom to give oxonium ions and halide ions:

$$HHal(g) + H_2O(l) \rightleftharpoons H_3O^+(aq) + Hal^-(aq) \tag{7.1}$$

In the cases of all the hydrogen halides except HF, the dissociation in aqueous solution is practically complete and they are strong acids. HF, with its tendency towards hydrogen bond formation, is a weak acid in aqueous solution, because of the equilibrium:

$$2HF(g) + H_2O(l) \rightleftharpoons H_3O^+(aq) + HF_2^-(aq) \qquad (7.2)$$

and because any fluoride ions are closely paired with oxonium ions: $H_3O^+F^-$.

Worked Problem 7.2

Q The ion HF_2^- and the unknown compound HeF_2 both have two electrons in the valence shell of the central atom. Why does the latter compound not exist?

A The ion HF_2^- has an energetic advantage over HeF_2 of being stabilized by hydration energy in aqueous solution and lattice energy in crystals. For bonding, the two electrons of the $1s^2$ configurations must be unpaired by promoting one of them to the next available orbital, 2s/2p. To achieve this in H^- would require the expenditure of $^3/_4 I_H$ (see equation 3.11), but in the He case the energy required would be $3I_H$ if all interelectronic repulsion terms were ignored. The promotion energy used in He would prevent the process from being viable for the formation of the compound.

7.2.2 Fluorides of the Elements of the Second Period

The elements of the second period are the least metallic in each group, and only lithium forms an ionic fluoride, LiF. Beryllium fluoride, BeF_2, exists in the solid state as chains of tetrahedrally coordinated beryllium atoms, with each pair of beryllium atoms bridged by two fluorine atoms, as shown in Figure 7.1 The chain structure may be thought of in terms of each beryllium atom forming two electron-pair bonds with two fluorine atoms, and then acquiring two more electron pairs as two more fluorine atoms (bonded to the two adjacent Be atoms) use their lone pairs of electrons to produce two coordinate bonds. In this way, the beryllium atom acquires its share of an octet of electrons associated with maximum stability.

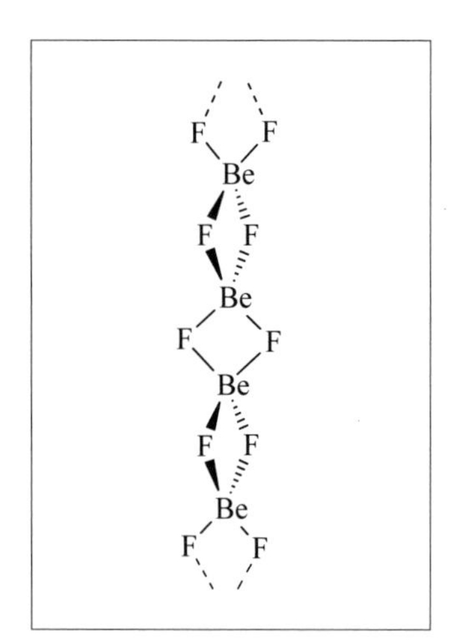

Figure 7.1 The bridged structure of BeF_2

Worked Problem 7.3

Q In the bridged structure of BeF_2, is the Be atom divalent or fourvalent?

A According to the definition of valency as the number of electron-pair bonds the element participates in, the Be atom is divalent. It does participate in four bonds, two of which are coordinate; in these the two electron pairs are supplied by the bridging fluorine atoms.

Boron forms the trifluoride molecule, BF_3, which is planar. If the boron atom simply exerted its trivalence, the B–F bonds would be straightforward single covalent bonds. The bonds are shorter than expected for single bonds, and the boron atom would not have a share in an octet of electrons. As in BeF_2, the deficiency of electrons is made up by the use of a lone pair of electrons from a fluorine atom, but the process is **intramolecular** rather than the **intermolecular bridging** in BeF_2. One fluorine atom donates an electron pair to the central boron atom to make up the octet. All three B–F bonds are identical, which is explained in terms of valence bond theory by the canonical forms shown in Figure 7.2. Molecular orbital theory retains the three single bonds, but has an extra pair of fluorine 2p electrons delocalized over the molecule evenly to give the same effect.

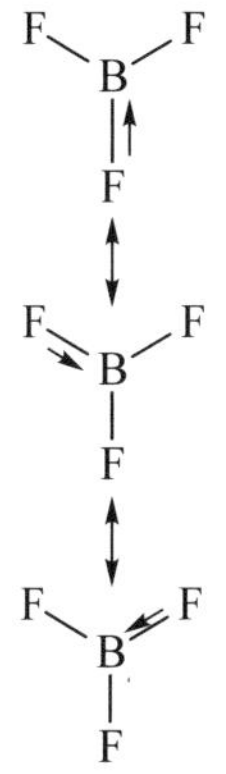

The other three elements of the second period form covalent molecular fluorides, CF_4, NF_3 and OF_2, in which the central atoms obey the octet rule and have their normal valencies.

Figure 7.2 The valence-bond canonical forms of the BF_3 molecule

Box 7.1 The +2 Oxidation State of Carbon

Although the molecule CF_2 does have a transient existence, there are stable compounds which have the same empirical formula but are very different in structure. The empirical unit CF_2 forms the compound C_2F_4 as a dimer. The dimer, tetrafluoroethene, is a stable molecule with the structure shown in Figure 7.3. This contains a C=C double bond, so that both carbon atoms have a share in an octet of valency electrons and are four-valent.

The ultimate product of the polymerization of the CF_2 unit is the solid known as poly(tetrafluoroethene) or PTFE (Teflon), used in non-stick surface coverings. In the polymer there are C–C linkages between the tetrafluoroethene monomer units. All the atoms obey the octet rule, all the carbon atoms are four-valent, and each of the fluorine atoms participate in one single bond. That the oxidation state of the carbon is +2 is very misleading.

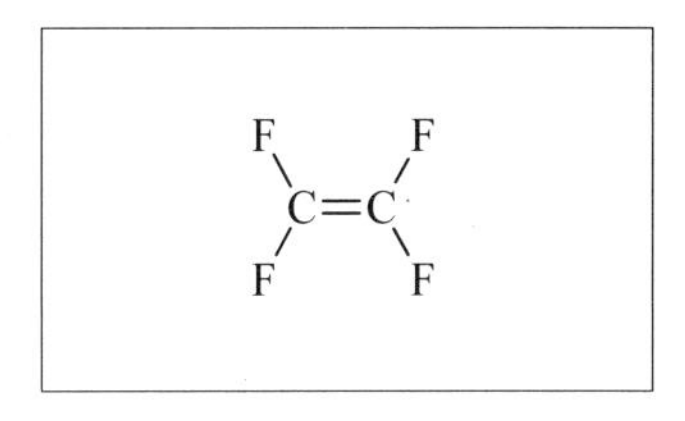

Figure 7.3 The structure of the C_2F_4 molecule

Across the second period the fluorides show the transition from ionic to covalent between Li and Be, and there is a transition from the chain structure of BeF_2 to molecular BF_3 as two examples of dealing with the electron deficiency of the non-metallic elements, and finally there are the molecular fluorides of C, N and O.

Worked Problem 7.4

Q Do the melting points of the fluorides of the second period indicate the type of bonding occurring? The melting points of the fluorides are:

LiF	*BeF$_2$*	*BF$_3$*	*CF$_4$*	*NF$_3$*	*OF$_2$*
848 °C	535 °C	–127 °C	–184 °C	–206 °C	–223 °C

A Yes. Comparisons of melting (and boiling) points can be a guide to changes in bonding; individual values are not so helpful and can be misleading. The value for LiF is that expected for an ionic compound, and the much lower value for BeF_2 indicates a high degree of covalency in the bridged polymeric solid. The very different and much lower values of the melting points of the fluorides of Groups 13, 14, 15 and 16 are consistent with their discrete molecular (*i.e.* covalent) forms.

7.2.3 Fluorides of the s- and p-Block Elements of the 3rd, 4th, 5th and 6th Periods

The fluorides of the s- and p-block elements of the periods beyond the second are summarized in Table 7.1. The oxidizing power of fluorine produces numerous examples of hypervalency, in which the maximum valency is shown by the most stable compounds. The only exception to this statement is provided by the bismuth fluorides: the hypervalent BiF_5 (linear polymer) is considerably less stable than the +3 compound. There are a few examples of the inert pair effect, but the fluorides with the group valency are the more stable.

The periodicity of the s- and p-block fluorides is dominated by patterns of ionic compounds in Groups 1 and 2, polymeric fluorine-bridged compounds in Groups 13–15, and then at the ends of each period there are molecular compounds. The ionic and molecular compounds are stable in those forms because there are sufficient electrons either to produce ions with Group 18 configurations, or to form sufficient covalent bonds with the available valence electrons. The intermediate compounds are stabilized in the condensed phases by the formation of fluorine bridges, in which fluorine atoms which are covalently bonded to one central element use one of their lone pairs of electrons to form a coordinate bond with another adjacent central element. This is shown in Figure 7.4.

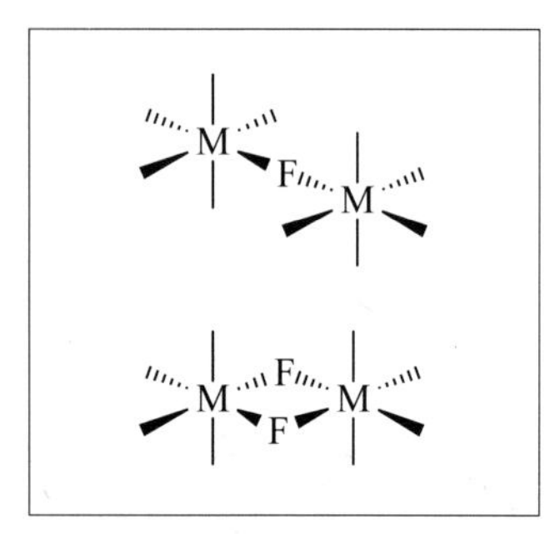

Figure 7.4 Examples of fluorine-bridged structures

Table 7.1 The fluorides of the s- and p-block elements of Periods 3–6. Compounds in which the central element exhibits its group valency, as expected from the octet theory, are shown in red and the more stable or most stable compound for each element is shown in **bold type** if there is a choice. The compounds shown with dark backgrounds are essentially ionic, those with the lighter backgrounds are polymeric with fluorine bridges linking the various units, and the remaining compounds are molecular.

Oxidation state	*Group* 1	2	13	14	15	16	17	18
VI						**SF_6**		
V					**PF_5**		**ClF_5**	
IV				SiF_4		SF_4		
III			AlF_3		PF_3		ClF_3	
II		MgF_2		SiF_2		SF_2		
I	NaF						ClF	
VI						**SeF_6**		
V					**AsF_5**		**BrF_5**	
IV				**GeF_4**		SeF_4		
III			GaF_3		AsF_3		BrF_3	
II		CaF_2		GeF_2		SeF_2		KrF_2
I	KF						BrF	
VII							**IF_7**	
VI						**TeF_6**		**XeF_6**
V					**SbF_5**		IF_5	
IV				**SnF_4**		TeF_4		XeF_4
III			InF_3		SbF_3		IF_3	
II		SrF_2		SnF_2				XeF_2
I	RbF							
V					BiF_5			
IV				**PbF_4**				
III			**TlF_3**		**BiF_3**			
II		BaF_2		PbF_2				
I	CsF		TlF					

The formation of fluorine-bridged compounds allows the central elements to achieve octets in the case of the Group 13 elements, and in all cases to achieve a degree of packing in the solid state that is maximized by the central element having a preferred coordination number of six. In some cases, however, the coordination is not regular, and the molecular units are identifiable by their smaller element–fluorine distances.

Worked Problem 7.5

Q Do the melting points of the s- and p-block (higher valency) fluorides of the third period indicate the type of bonding occurring? The melting points of the fluorides are:

KF	CaF_2	GaF_3[a]	GeF_4	AsF_5	SeF_6
857 °C	1423 °C	800 °C	–37 °C	–63 °C	–40 °C

[a]This compound sublimes at the temperature given.

A Yes. The high values for KF and CaF_2 are those expected for ionic compounds. The increase in m.p. from KF to CaF_2 is expected from the doubly charged Ca^{2+} ion and its effect on the lattice enthalpy of the difluoride. The lower value for GaF_3 is an indication of the onset of covalency in the polymeric structure. That the compound sublimes instead of melting is another indication of its polymeric nature, the bridging between units persisting in the vapour phase, Ga_2F_6. The sharp fall in melting points in the other fluorides marks the transition to discrete molecular (*i.e.* covalent) forms.

7.2.4 Fluorides of the Transition Elements

The oxidation states of the fluorides formed by the three series of transition elements are shown in Table 7.2. At the left-hand side of each period the most stable oxidation states are those corresponding to the use of all the valence electrons of the transition element atoms. The fluorides of the elements in oxidation states lower than their maximum are less stable.

In the first transition series there is a dominance of the trifluorides from Cr to Co, and from Ni to Zn only the difluorides exist. This is consistent with the increasing difficulty of oxidizing these elements and the accompanying instability of high coordination numbers around relatively small central metal atoms. These factors are less important with the second and third series of transition elements, which form a greater range of fluorides with higher oxidation states having greater thermodynamic stability than their counterparts in the groups of the first series.

The nature of the bonding of the fluorides varies with oxidation state and two sets are described as examples. The fluorides of vanadium and ruthenium are chosen because there are four in each set and they are all well characterized. The fluorides of vanadium vary from the ionic VF_2, through the polymeric VF_3 and VF_4, to VF_5 which is polymeric in the

Table 7.2 The oxidation states of the fluorides of the transition elements. The numbers also indicate the stoichiometries of the compounds formed, MF_x. Those shown in red are the most thermodynamically stable, as far as data are available

Group									
3	*4*	*5*	*6*	*7*	*8*	*9*	*10*	*11*	*12*
Sc	*Ti*	*V*	*Cr*	*Mn*	*Fe*	*Co*	*Ni*	*Cu*	*Zn*
		+5	+5						
	+4	+4	+4	+4			+4		
+3	+3	+3	+3	+3	+3	+3	+3		
		+2	+2	+2	+2	+2	+2	+2	+2
Y	*Zr*	*Nb*	*Mo*	*Tc*	*Ru*	*Rh*	*Pd*	*Ag*	*Cd*
			+6	+6	+6	+6			
		+5	+5	+5	+5	+5			
	+4	+4	+4		+4	+4	+4		
+3	+3	+3	+3		+3	+3		+3	
							+2	+2	+2
								+1	
Lu	*Hf*	*Ta*	*W*	*Re*	*Os*	*Ir*	*Pt*	*Au*	*Hg*
				+7					
			+6	+6	+6	+6	+6		
		+5	+5	+5	+5	+5	+5	+5	
	+4		+4	+4	+4	+4	+4		
+3						+3		+3	
									+2
									+1

solid state but is molecular in the gas phase. Those of ruthenium vary in their modes of polymerization from RuF_3, in which the Ru atoms are surrounded by six fluorine atoms, three from adjoining RuF_3 units; RuF_4 which has a polymeric structure in which two fluorine atoms are shared by each pair of adjacent Ru atoms; RuF_5, which is a tetramer in the solid state (*i.e.* four units linked together by Ru–F–Ru bridges); and RuF_6, which is molecular. All the ruthenium fluorides are classed as covalent; only VF_2 of the vanadium fluorides is ionic, because of the low oxidation state of the vanadium. The transition from ionic, through polymeric covalent to covalent molecular compounds is understandable in terms of the increasing difficulty of successive ionizations which limits the ionic compounds to lower oxidation states, and the 18-electron rule which determines the physical state of the covalent compounds. The details of the bonding in transition metal compounds are left to other books in this series, but a brief outline of the theory is sufficient for the present discussion. The valency orbitals of the transition elements are the five

The $3d_{z^2}$ and $3d_{x^2-y^2}$ orbitals contribute to σ-bonding and anti-bonding orbitals, the $3d_{xy}$, $3d_{xz}$ and $3d_{yz}$ orbitals contribute to π-bonding and anti-bonding orbitals in Figure 7.5. Because the overlap of σ orbitals is more efficient than that when π orbitals overlap, there is an energy gap between the σ and π bonding molecular orbitals, with the σ orbitals more stable than the π. The reverse is the case with the anti-bonding orbitals, with the π* orbitals more stable than the σ*.

(n – 1)d, the single ns and the three np atomic orbitals. These nine atomic orbitals have the correct symmetries to overlap with groups of six octahedrally positioned ligand orbitals to produce nine bonding orbitals (six σ-type and three π-type) and nine anti-bonding orbitals. In addition, there are nine non-bonding orbitals which are restricted to the ligand groups. A simplified diagram of the energies of these orbitals is shown in Figure 7.5.

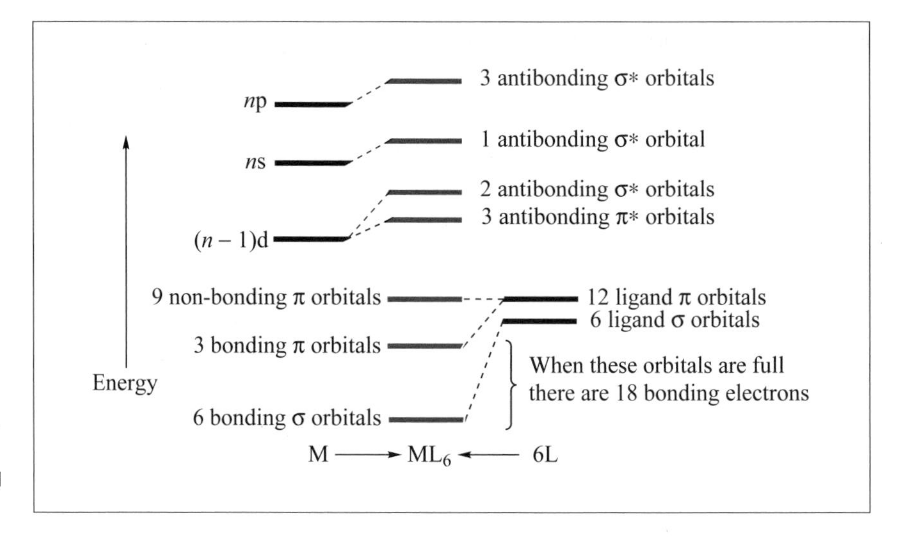

Figure 7.5 The molecular orbitals of an octahedrally coordinated transition metal atom. The anti-bonding orbitals are indicated by the asterisks

Octahedral coordination is very common in transition element chemistry, and is the basis of the rationalization of the bonding of their compounds. If a compound is formed in which the nine bonding orbitals are filled, the 18-electron rule is satisfied and the central transition metal atom has its share in 18 electrons. That would be the case for MoF_6, in which the molybdenum atom is in its formal +6 oxidation state, corresponding to the complete removal of its six valence electrons which are transferred to the six fluorine atoms to produce six fluoride ions. If the compound is regarded initially as Mo^{6+} + $6F^-$, the ligand fluoride ions may be considered to donate six pairs of σ-type electrons to the appropriate bonding orbitals of Figure 7.5 and to donate three pairs of π-type orbitals to the three π-bonding orbitals. That is sufficient bonding to make the compound stable as a molecular compound, melting at 17.5 °C. Molybdenum also forms a +5 fluoride; this could form a molecule which has a trigonal-bipyramidal shape, but a more stable arrangement is a small polymer with four monomer units, $(MoF_5)_4$, in which each pair of monomer units shares a fluorine atom forming an Mo–F–Mo bridge as shown in Figure 7.6.

Small polymers with definite numbers of units are called **oligomers**.

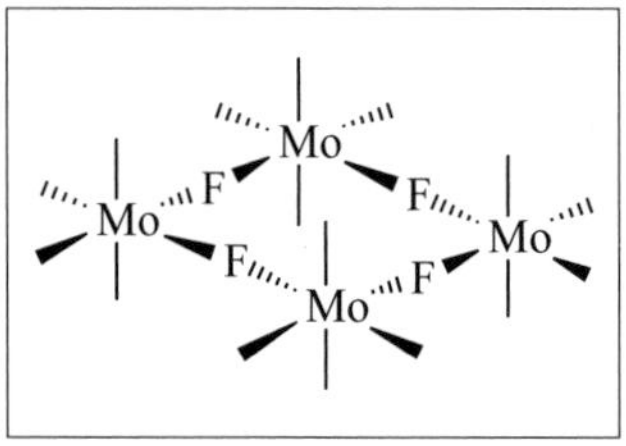

Figure 7.6 The structure of the $(MoF_5)_4$ oligomer

The Mo^V oligomer has four Mo atoms which are octahedrally coordinated. This allows each Mo atom to have its nine bonding orbitals filled, thus conferring the maximum stability on the metal centre. Each

Mo still possesses a d^1 single electron which occupies an anti-bonding orbital, but the destabilization of the oligomer is outweighed by the stability achieved by having octahedral coordination.

In a similar manner, fluorides with the stoichiometry MF_4 would have the capacity to make two fluoride bridges per metal atom, and those with the stoichiometry MF_3 could form three fluoride bridges per metal atom. These possibilities allow the formation of two-dimensional infinite polymer sheets and three-dimensional infinite polymer arrays, respectively, for the two stoichiometries. In addition to the polymeric possibilities, there are many examples of the excess of d electrons on each metal atom actually forming direct metal–metal bonds. This can lead to a delocalization of the metal d electrons that allows the compounds to be electrical conductors.

The general variability of the oxidation states of the transition elements is explicable in terms of the closeness of values of successive ionization energies and, for any coordination number, the existence of odd numbers of anti-bonding electrons is not unusual. Such anti-bonding electrons possess almost the same energies that they do in the isolated oxidation state, and do not destabilize the systems sufficiently to rule out their existence. In the latter halves of the transition series, successive ionization energies increase and the attainment of higher oxidation states is less apparent.

Worked Problem 7.6

Q Comment on the relationship between the melting points of the fluorides of chromium and the bonding in their structures.

CrF_2	CrF_3	CrF_4[a]	CrF_5	CrF_6[b]
894 °C	1404 °C	277 °C	34 °C	–100 °C

[a] Sublimation temperature. [b] Decomposes above this temperature.

A The melting point data are consistent with the transition from ionic +2 and +3 compounds through the polymeric covalent +4 compound to the discrete molecular +5 and +6 compounds, the latter being unstable owing to the steric difficulties of arranging six fluorine atoms around a relatively small metal atom.

7.2.5 Fluorides of the f-Block Elements

Lanthanide Fluorides

The lanthanide elements all form +3 fluorides, which are the more stable except in the case of cerium for which the +4 fluoride is the more stable. Samarium and europium form +2 fluorides, and in addition to cerium, Pr and Tb form +4 fluorides. The +2 fluorides are ionic, but the others are better described as covalent polymeric structures. The +3 fluorides have central metal atoms surrounded by nine fluorine atoms with two more fluorine atoms not far away. The compounds are good conductors of electricity, and this property may be understood if the lanthanides form +3 ions with one electron per atom occupying a conduction band. This would be possible if a 4f electron were to be promoted to the 3d conduction band, the 4f orbitals being too compact to form bands by overlapping throughout the structure. A coordination number of nine closest neighbours is a common characteristic of lanthanide chemistry.

Actinide Fluorides

The actinides offer a greater range of oxidation states in their fluorides than do the lanthanides, particularly in the first half of the series. The oxidation states of the actinide fluorides are shown in Table 7.3. The pattern of stable oxidation states is very much like those of the transition elements rather than like the lanthanides, but towards the half-full 5f orbitals stage and beyond the actinides tend to mirror the lanthanides. The trans-californium elements are not included in Table 7.3. They are more lanthanide-like and form only +3 fluorides.

Table 7.3 Oxidation states of the fluorides of the actinides

Ac	*Th*	*Pa*	*U*	*Np*	*Pu*	*Am*	*Cm*	*Bk*	*Cf*
			+6	+6	+6				
		+5	+5	+5					
	+4	+4	+4	+4	+4	+4	+4	+4	+4
+3			+3	+3	+3	+3	+3	+3	+3

7.3 Oxides of the Elements

Although oxygen is not potentially as oxidizing as fluorine, the element is a powerful oxidant and combines with almost all the elements to form oxides, compounds containing formally the oxide ion O^{2-}. In some cases the element is a better oxidant than fluorine as it can form double bonds with the central element of a compound and does not suffer from the steric difficulties experienced by fluorine (see Section 7.1).

Box 7.2 Forms of Oxygen in Compounds

Oxygen exists as the diatomic molecule dioxygen, in which there is a "double bond" formed from the eight 2p electrons of the two oxygen atoms. The simplified diagram of Figure 7.7 shows the molecular orbitals formed from the σ overlap of the two 2p orbitals that lie along the molecular axis and the π overlap of the other two sets of 2p orbitals that are perpendicular to the molecular axis.

Magnetic measurements show that the dioxygen molecule does possess two unpaired electrons.

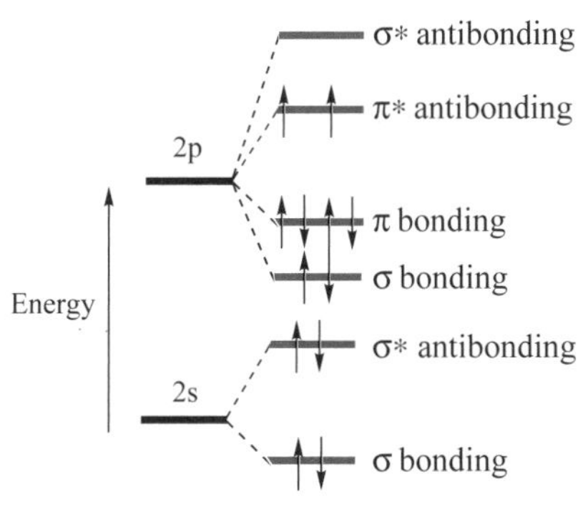

Figure 7.7 A molecular orbital diagram for the dioxygen molecule

There are three bonding orbitals and three anti-bonding orbitals, and they are filled according to the *aufbau* principle. The three bonding orbitals are filled by six of the 2p valency electrons, and the other two valency electrons occupy singly the two anti-bonding π* orbitals. The anti-bonding electrons (two) cancel the bonding effect of two of the electrons in the bonding orbitals so that the overall bonding is due to the other two bonding electron pairs.

The two pairs of 2s electrons occupy σ bonding and σ* anti-bonding orbitals, and do not significantly contribute to the bonding.

This rather complicated double bond causes the dioxygen molecule to have a dissociation enthalpy of 496 kJ mol^{-1}, which is a relatively strong bond. The reactivity of dioxygen does not always depend on the dissociation of the molecule. It already has two unpaired electrons in its anti-bonding π orbitals ready to form electron pair bonds or to engage in the formation of ions. The various ways in which dioxygen can react are:

- *Ion formation*
  (i) The molecule can be **oxidized** to O_2^+ by losing an electron. The compound PtF_6 is a very powerful oxidizing agent, and reacts with dioxygen to give the ionic compound $O_2^+PtF_6^-$.
  (ii) The molecule can undergo one-electron reduction to form the superoxide ion, O_2^-. When caesium metal burns in dioxygen the product is the superoxide, CsO_2.

(iii) The molecule can undergo a two-electron reduction to form the peroxide ion, O_2^{2-}. When metallic sodium burns in dioxygen the main product is the peroxide, Na_2O_2.
(iv) The molecule can undergo a four-electron reduction to form two oxide ions, $2O^{2-}$. In this case the molecule is fully dissociated. Oxide formation takes place in the reactions of most of the metallic elements with dioxygen.

- *Covalent bond formation*
  With the less electropositive elements, dioxygen reacts to give covalent oxides, but there are many peroxides, *e.g.* H_2O_2 in which the O–O bond is retained.

The discussion is divided into sections dealing with the s- and p-block elements, the transition elements and the lanthanide and actinide elements. In addition to the stoichiometry and bonding of the oxides, their reactions with water are described and whether the products of such reactions are acidic or alkaline.

7.3.1 Acidic and Basic Oxides

In the aqueous system, the **Brønsted–Lowry definitions** of acids and bases are of general applicability. A **Brønsted acid** is a proton donor. Acidic oxides react with water to produce **oxonium ions** in solution, *e.g.*

$$N_2O_5(g) + H_2O(l) \rightarrow 2HNO_3(aq) \tag{7.3}$$

$$HNO_3(aq) + H_2O(l) \rightleftharpoons H_3O^+(aq) + NO_3^- \tag{7.4}$$

The other well-used definitions of acidic and basic behaviour are those of G. N. Lewis. He proposed that an acid is an **electron-pair acceptor** and a base is an **electron-pair donor**. The Lewis definitions are not used in this text, but many metal fluorides, for example, behave as **Lewis acids**, *e.g.* $SbF_5 + F^- \rightarrow SbF_6^-$. In that case, the fluoride ion acts as a **Lewis base** and donates an electron pair to the Sb centre.

In the second reaction the water molecule acts as a base, accepting a proton to give the oxonium ion.

A **Brønsted base** is a proton acceptor. Basic oxides react with water to produce **hydroxide ions** in solution, *e.g.*

$$Na_2O(s) + H_2O(l) \rightarrow 2Na^+(aq) + 2OH^-(aq) \tag{7.5}$$

In effect the oxide ion acts as a base, accepting a proton from the water molecule to produce a hydroxide ion:

$$O^{2-} + H_2O(l) \rightarrow 2OH^-(aq) \tag{7.6}$$

The two ions produced by the action of water as either a base or an acid are those which are produced by the **auto-ionization** of the solvent, water: $2H_2O(l) \rightleftharpoons H_3O^+(aq) + OH^-(aq)$ and the two ions represent the strongest acid [$H_3O^+(aq)$] and the strongest base [$OH^-(aq)$] that can exist in aqueous solution.

Oxides are best described as either acidic or basic depending whether they dissolve in water to give either $H_3O^+(aq)$ or $OH^-(aq)$, respectively. The factors that govern whether a compound is acidic or basic can be outlined in terms of the possible reactions of *hypothetical* hydroxides.

Consider one such hydroxide, which has been produced by the addition of a water molecule to an oxide of an element in its +1 oxidation state:

$$E_2O + H_2O \rightarrow 2EOH \tag{7.7}$$

A hypothetical **dissociation** of the compound EOH can be allowed, giving three fragments: $E^+ + O^{2-} + H^+$. This arranges a competition for the doubly negative oxide ion between the proton and the ion E^+, the outcome of which will depend on the electrostatic attractions and the subsequent hydration enthalpies of the two ions produced. The two possibilities are: $EO^-(aq) + H^+(aq)$ if the attraction between E^+ and O^{2-} is dominant, and $E^+(aq) + OH^-(aq)$ if the proton has the greater claim on the oxide ion. In both cases the added contributions to the exothermicity of the overall outcome from the enthalpies of hydration of the ions are important. If $EO^-(aq) + H^+(aq)$ are the energetically preferred products, the oxide is called acidic. If $E^+(aq) + OH^-(aq)$ are energetically preferred, the oxide is called basic.

Because the proton is considerably smaller than any possible E^+ ions, the outcome of the modelling is that E_2O would be expected to be basic in character. Furthermore, in a comparison of basic oxides, the one with the largest E^+ ion would be expected to be the most basic, *i.e.* have the highest exothermicity for its dissociation giving hydroxide ions. The only way of reducing the efficiency of the basic dissociation of a hydroxide is to increase the oxidation state of the element E.

An extreme case is that for the oxide E_2O_7 with the element E in its +7 oxidation state. Such an oxide could react with water to give the undissociated compound:

$$E_2O_7 + H_2O \rightarrow 2O_3EOH \tag{7.8}$$

The neutral hydroxide can then be dissociated to give the three ions $[O_3E^{VII}]^+ + O^{2-} + H^+$, and the two positive ions compete for the oxide ion. Since the central element is in its +7 oxidation state it would have a very great attraction for another oxide ion, and the more likely outcome of the competition is that the proton will lose out and the dissociation products of the neutral compound will be to give the ions $[O_4E^{VII}]^-(aq) + H^+(aq)$.

The results of the above hypothesizing may be generalized:

(i) **Basic oxides** are those in which the element is in a *low oxidation state* and that positive ions from *large atoms* would favour *basicity*, rather than those from small elements. Most basic oxides dissolve in water to give alkaline solutions. They, and those that are insoluble in water, react to neutralize acids.

(ii) **Acidic oxides** are those in which the element is in a *high oxidation* *state*; oxides of the elements with *small atoms* would favour *acidity* rather than those with large atoms. Acidic oxides dissolve in water to give acid solutions or, if they are insoluble in water, neutralize alkaline solutions.

Intermediate results are possible for the two cases. Where both dissociations are feasible, such compounds are termed **amphoteric oxides**.

7.3.2 Water

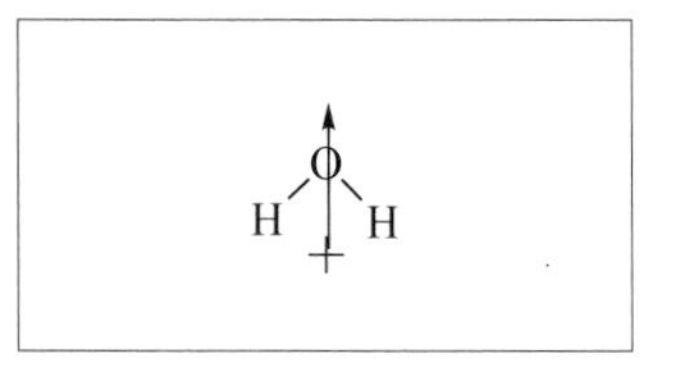

Figure 7.8 A diagram of the water molecule showing its dipole

The oxide of hydrogen, H_2O is a bent (104.5°) covalent molecule with polarized bonds (*i.e.* the electron pairs are shared unequally by the participating atoms) such that there is a **permanent dipole moment** as indicated in Figure 7.8. The oxygen atom has a partial negative charge that is balanced by partial positive charges on the two hydrogen atoms. From its low RMM value it might be expected to be a gas at 298 K, but because of considerable **hydrogen bonding** it is a liquid.

7.3.3 Oxides of the Second Period

The main oxides of the elements of the second period are shown in Table 7.4. There is the expected pattern, but there are exceptions which are dealt with below. Across the period it would be expected that the compounds would be less ionic/more covalent as the differences in values of electronegativity coefficients between the elements and oxygen decrease. Lithium oxide is ionic and dissolves in water to give a solution of lithium hydroxide. It is a basic oxide. The ionic/covalent transition occurs very early, and BeO is a very high-melting (2530 °C) covalent oxide. The doubly charged Be^{2+} ion would be very small, and the application of Fajans' rules would imply that its compounds should be covalent. The oxide does not dissolve in water, but reacts with concentrated sulfuric

Table 7.4 Oxides of the elements of the second period. The compounds with the "octet" group oxidation states/valencies are shown in red

Oxidation state	*Group*							
	1	*2*	*13*	*14*	*15*	*16*	*17*	*18*
	Li	*Be*	*B*	*C*	*N*	*O*	*F*	*Ne*
V					N_2O_5			
IV				CO_2	N_2O_4, NO_2			
III			B_2O_3		N_2O_3			
II		BeO		CO	NO			
I	Li_2O				N_2O			
−1							OF_2	

acid to give the Be^{2+}(aq) ion and with sodium hydroxide to give the beryllate ion, best written as $[Be(OH)_4]^-$. BeO is therefore an amphoteric oxide.

Boron forms B_2O_3, which is a covalent compound of varied structure. It normally has a glass-type randomly oriented structure of B_3O_3 rings with bridging oxygen atoms. This property makes the compound useful in the production of borosilicate glass (*e.g.* Pyrex). The oxide dissolves in acid to form borates; the simplest ion is BO_3^{3-} (isoelectronic with BF_3) and there are many polymeric varieties.

Carbon dioxide is a covalent linear triatomic molecule and is the anhydride of carbonic acid, H_2CO_3, which is the parent acid of many hydrogencarbonate (HCO_3^-) and carbonate (CO_3^{2-}) salts.

With nitrogen the "octet" compound is N_2O_3, but the atoms are *not* bonded with conventional single and double bonds, O=N–O–N=O, as might be expected with both atoms exerting their group valencies. The actual structure is shown in Figure 7.9.

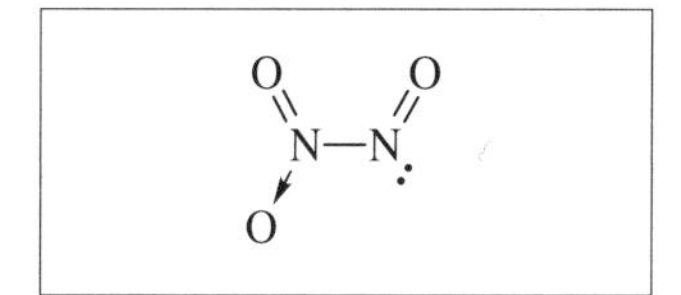

Figure 7.9 The structure of the N_2O_3 molecule

The molecule is planar and has an N–N bond. The nitrogen bonded to the two oxygen atoms must alternate its bonding to those atoms between double and coordinate linkages. Molecular orbital theory explains the planarity by having 2p orbitals on all five atoms entering into π-type delocalized bonding. The compound dissociates into NO + NO_2 in the gas phase (above 3.5 °C). It dissolves in water to give a solution of nitrous acid [nitric(III) acid]:

$$N_2O_3 + H_2O(l) \rightarrow 2HON{=}O(aq) \quad (7.9)$$

which is unstable above 5 °C but is the parent acid to many nitrite [nitrate(III) ion] salts.

The final oxide of the second period is OF_2, possibly misrepresented in Table 7.4 as a +1 state of fluorine. This is another example of the oxidation state concept misleading the reader, and the compound is better thought of as *oxygen difluoride* with the oxygen atom having a partial positive charge. It would be too extreme to regard the compound as one of oxygen(II). The compound is molecular and not very stable, but the octet rule is obeyed. It dissolves in water to give a neutral solution which decomposes slowly, more rapidly with a base:

$$F_2O(g) + 2OH^-(aq) \rightarrow 2F^-(aq) + O_2(g) + H_2O(l) \quad (7.10)$$

It is of interest to note that the fluorine atoms get the electrons!

Worked Problem 7.7

Q Do the melting points of the oxides of the second period show similar variations to those of the corresponding fluorides (see the

Worked Problem 7.4 in Section 7.2.2) of the elements? The melting points of some oxides of the second period are:

Li_2O	BeO	B_2O_3	CO_2[a]	N_2O_3	OF_2
>1700 °C	2530 °C	45 °C	–56.6 °C	–102 °C	–223.8 °C

[a] Under 5 atm pressure.

A Yes, but some differences of interpretation are necessary. Lithium oxide is ionic, but BeO is best regarded as a giant covalent array (it has the wurtzite, ZnS, structure). The Be^{2+} ion has only been characterized in aqueous solution, Be^{2+}(aq) or $[Be(H_2O)_4]^{2+}$, and in hydrated compounds, *e.g.* $[Be(H_2O)_4]Cl_2$. Although boron oxide does not consist of discrete molecules, it has a low melting point associated with its "glassy" state. The other oxides have the low melting points expected from covalent discrete molecules. Liquid CO_2 only exists under a 5 atm pressure of the gas, but is a very good solvent for non-polar molecules, *e.g.* it dissolves caffeine from coffee beans.

The oxides of the second period have very different thermodynamic stabilities, as indicated by the standard enthalpies of formation given in Table 7.5. The first four oxides of Table 7.5 are thermodynamically very stable *with respect to their constituent elements*, but the oxides of nitrogen and fluorine are **endothermic**, *i.e.* they are *less stable* than their constituent elements. The stable oxides derive their stabilities from their bonding, ionic or covalent, with extra contributions from the solid-state aggregation. Dinitrogen trioxide is less stable than N_2 + $^3/_2O_2$, the reason being connected with the high values of the bond dissociation energies of those molecules. The dioxygen stability contributes to the endothermicity of OF_2.

Table 7.5 Standard enthalpies of formation for the second period oxides

Oxide	$\Delta_f H^\circ$ *(298 K)/kJ mol^{-1}*
Li_2O(s)	–596
BeO(s)	–611
B_2O_3(s)	–1264
CO_2(g)	–394
N_2O_3(g)	93
F_2O(g)	23

Apparent Exceptions to the Octet Rule

There are apparent exceptions to the octet rule in Group 14 (carbon monoxide) and Group 15 (the oxides of nitrogen other than N_2O_3. Carbon monoxide has carbon in its formal oxidation state of +2. Carbon monoxide has four valency electrons ($2s^22p^2$) and oxygen has six ($2s^22p^4$) and the 10 electrons occupy the σ and π molecular orbitals shown in Figure 7.10.

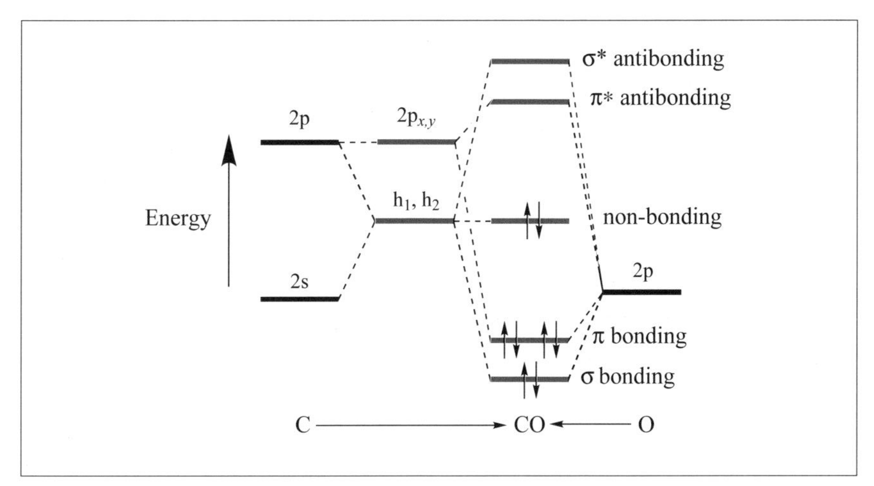

Figure 7.10 A molecular orbital diagram for the CO molecule. The two orbitals labelled h_1 and h_2 are hybrid mixtures of the 2s and the $2p_z$ carbon orbitals. They lie along the molecular axis in diametrically opposite directions. One of the hybrid orbitals is used to make a σ-bond with the oxygen atom, the other contains a lone pair of electrons and is directed away from the bonding region. The 2s orbital of the oxygen atom is ignored

There are six bonding electrons which represent a triple bond in the compound. The very strong bond (dissociation enthalpy = 1090 kJ mol^{-1}) is responsible for the general stability of CO. It does not react with water as many oxides do. Theoretically, carbon monoxide is the anhydride of formic acid, HCOOH, and may be produced by the dehydration of that acid by concentrated sulfuric acid. To regard the carbon as having an oxidation state of +2 is misleading. To a first approximation, both atoms in CO have their "octet" share of eight electrons. They atoms share the six bonding electrons and they each have a non-bonding pair to satisfy the rule.

The other oxides of nitrogen also provide examples of the breakdown of the usefulness of the oxidation state concept and its potential misleading interpretations. Dinitrogen monoxide, N_2O, is N^I if the oxygen atom is in its –2 state by convention. A simple valence bond view of the molecule is to write its structure with an N≡N triple bond, with one nitrogen using a lone pair of electrons to form a coordinate bond with the oxygen atom: N≡N→O. Or it could be written with two double bonds, if the central nitrogen atom loses an electron to the terminal nitrogen atom: $^-N{=}N^+{=}O$. In that case the central nitrogen, N^+, is behaving like a carbon atom and the terminal nitrogen, N^-, is behaving like an oxygen atom. The bonding in the molecule is very similar to that in CO_2, the two molecules being linear and isoelectronic, and there is no exception to the octet rule. Dinitrogen monoxide does not dissolve in water, but it is in theory the anhydride of hyponitrous acid, HON=NOH. The acid can be produced in solution by the reactions:

$$Na_2O + N_2O \rightarrow Na_2(ON{=}NO) \tag{7.11}$$

$$Na_2(ON{=}NO) + 2H_3O^+(aq) \rightarrow 2Na^+(aq) + HON{=}NOH + 2H_2O(l) \tag{7.12}$$

Nitrogen monoxide, NO, is *not* an example of N^{II}; the difference in electronegativity coefficients between N and O is only 0.5 and little charge separation is expected in the molecule. Its bonding is understandable from Figure 7.10, as it has one more electron than CO. The extra electron occupies a π anti-bonding orbital and gives NO its free-radical character. The presence of the anti-bonding electron reduces the bond dissociation enthalpy relative to that of CO. In the solid state the molecule dimerizes. NO reacts rapidly with dioxygen to give NO_2, but it is not an acid anhydride.

Free radicals are molecules or fragments of molecules with usually one unpaired electron. They are generally unstable and very reactive. Free radicals are rarely stable; NO is one and dioxygen may be regarded as a *diradical* in that it has two unpaired electrons.

Dinitrogen tetroxide is the dimer of the free radical nitrogen dioxide. Both are formally compounds of N^{IV}. NO_2 may be written as in Figure 7.11, in which it retains a lone pair of electrons, shares two more electrons with the doubly bonded oxygen atom, and uses the remaining valence electron to form a single bond to the other oxygen atom, the latter retaining the "odd" electron which makes the molecule a free radical. Formulated in this manner the nitrogen obeys the octet rule, but the singly bonded oxygen has a share in only seven valence electrons. The dimer is formed via an N–N bond and may be formulated as in Figure 7.12 with both nitrogen atoms engaging in one single, one double and one coordinate bond, so that only four atomic orbitals of the nitrogen atoms are used and all the atoms obey the octet rule.

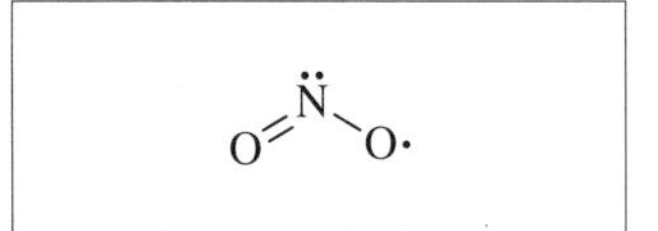

Figure 7.11 The structure of NO_2

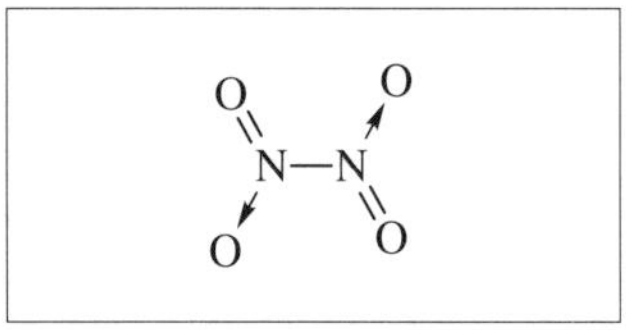

Figure 7.12 The structure of N_2O_4

The oxide dissolves in cold water to give a mixture of nitrous and nitric acids:

$$N_2O_4(g) + 2H_2O(l) \rightarrow HNO_2(aq) + H_3O^+(aq) + NO_3^-(aq) \tag{7.13}$$

Nitric acid is fully dissociated in aqueous solution.

Dinitrogen pentoxide is formally N^V, but the molecule may be formulated as in Figure 7.13, in which there is an N–O–N group and the two nitrogen atoms both form one double bond and one coordinate bond to two terminal oxygen atoms. All the atoms have their shared octet. The oxide is the anhydride of nitric acid:

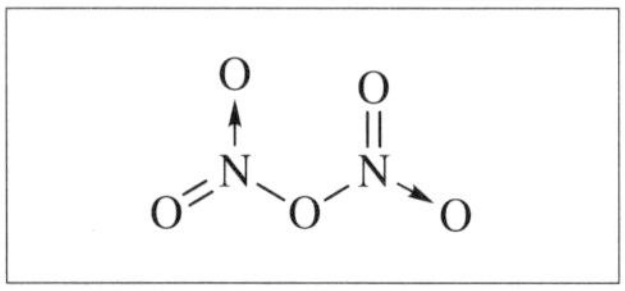

Figure 7.13 The structure of N_2O_5

$$N_2O_5(g) + 3H_2O(l) \rightarrow 2H_3O^+(aq) + 2NO_3^-(aq) \tag{7.14}$$

The poorly characterized nitrogen trioxide is formally a compound of N^{VI}, which is impossible given that the nitrogen atom only has five valence electrons. It may be formulated as in Figure 7.14, in which the central nitrogen atom makes a single bond to an oxygen which has the "odd" electron, a double bond to a second oxygen atom and a coordinate bond to the third oxygen atom. In that case the nitrogen atom has its octet share; only one of the oxygen atoms loses out, with a share in seven electrons.

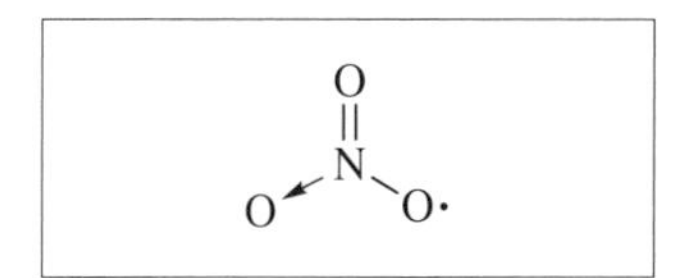

Figure 7.14 The structure of NO_3

7.3.4 Oxides of the s- and p-Blocks of the 3rd, 4th, 5th and 6th Periods

Table 7.6 lists the oxides of the elements of the s- and p-blocks of periods 3–6. There is a general pattern across each period. This is a transition from ionic basic oxides, through polymeric covalent oxides, some being amphoteric and the later ones being acidic, to the molecular acidic oxides of the later groups. Down each group there is a tendency for the oxides to be of similar stoichiometry and to be more basic, less acidic, towards the heavier members.

There are exceptions to the stoichiometry patterns, with some mixed oxides (*e.g.* Pb_3O_4) and, in particular with the Group 17 oxides, some compounds which seem out of place (see text). This latter point is because of the misleading ideas that can occur when using the oxidation state concept.

Dichlorine heptaoxide represents the highest oxidation state of chlorine and is a molecular compound with a Cl–O–Cl bridge. Dichlorine hexaoxide, formally a compound of Cl^{VI}, contains the chlorate(VII) ion, $ClO_2^+ClO_4^-$, with chlorine atoms in their +5 and +7 oxidation states. Chlorine dioxide is a free radical with 19 valency electrons and is very reactive. It is a mild oxidizing agent, used on a commercial scale for bleaching flour and wood pulp. It reacts with an alkaline solution to give a mixture of ClO_2^- and ClO_3^- ions. Its theoretical dimer, Cl_2O_4, a formal compound of Cl^{IV}, is another chlorate(VII), $Cl^+ClO_4^-$, and is very unstable. Diiodine tetraoxide, I_2O_4, has a complex structure which has polymeric chains of alternating iodine and oxygen atoms with the iodine atoms in adjacent chains "cross-linked" with IO_3 groups. Tetraiodine nonaoxide, I_4O_9, has a polymeric form of iodine-atom chains cross-linked with IO_3 groups. Regarded as $I^{3+}(IO_3^-)_3$ it may be thought of as a compound of iodine(III) and iodine(VII).

Table 7.6 The oxides of the s- and p-block elements of Periods 3–6. The oxides shown in red are the most stable, those with the darker background are ionic, those with the lighter background are polymeric and the remainder, without background, are molecular. The oxides with double stars are basic, those with one star are amphoteric and the starless remainder are acidic

Group 1	2	13	14	15	16	17	18
Na	*Mg*	*Al*	*Si*	*P*	*S*	*Cl*	*Ar*
Na_2O^{**}	MgO^{**}	$Al_2O_3^{*}$	SiO_2	P_4O_{10} P_4O_6	SO_3 SO_2	Cl_2O_7 Cl_2O_6 ClO_2/Cl_2O_4 Cl_2O_3 Cl_2O	
K	*Ca*	*Ga*	*Ge*	*As*	*Se*	*Br*	*Kr*
K_2O^{**}	CaO^{**}	$Ga_2O_3^{*}$ Ga_2O^{*}	GeO_2^{*} GeO^{*}	As_2O_5 As_2O_3	SeO_3 SeO_2	Br_2O_5 Br_2O	
Rb	*Sr*	*In*	*Sn*	*Sb*	*Te*	*I*	*Xe*
Rb_2O^{**}	SrO^{**}	$In_2O_3^{**}$ In_2O^{**}	SnO_2^{*} SnO^{*}	Sb_2O_5 Sb_2O_4 Sb_4O_6	TeO_3 Te_2O_5 TeO_2	I_2O_5 I_2O_4 I_4O_9	XeO_4 XeO_3
Cs	*Ba*	*Tl*	*Pb*	*Bi*	*Po*	*At*	*Rn*
Cs_2O^{**}	BaO^{**}	$Tl_2O_3^{**}$ Tl_2O^{**}	PbO_2^{*} $Pb_3O_4^{*}$ PbO^{*}	$Bi_2O_3^{*}$	PoO_2		

Worked Problem 7.8

Q Interpret the following melting points of the oxides of the third period in terms of their structures and bonding.

Na_2O^a	MgO	Al_2O_3	SiO_2
1275 °C	2852 °C	2072 °C	1723 °C
P_4O_6/P_4O_{10}	SO_2/SO_3	Cl_2O	
23.8/580 °C	−72.7/16.8 °C	−20 °C	

[a] Sublimes.

A The first three oxides are ionic, and although the triple charge on Al^{3+} would introduce some covalent character, Al_2O_3 does conduct electricity poorly in the molten state. [In the electrolytic manufacture of the metal the conductivity is increased by the addition of cryolite, Na_3AlF_6, which also reduces the melting point.] This shows that Al_2O_3 is by no means purely ionic. SiO_2 forms a covalent infinite array and has a correspondingly high melting point The other oxides have the low melting points expected from covalent discrete molecules, except for the P^V oxide which has the structure shown in Figure 7.15. The structure of the P^{III} oxide has the same basic cluster as the P^V oxide, but without the four terminal oxygen atoms.

The molecular compound has a melting point of 420 °C, but after heat treatment for some time the eventual most stable product is polymeric and has a melting point of 580 °C.

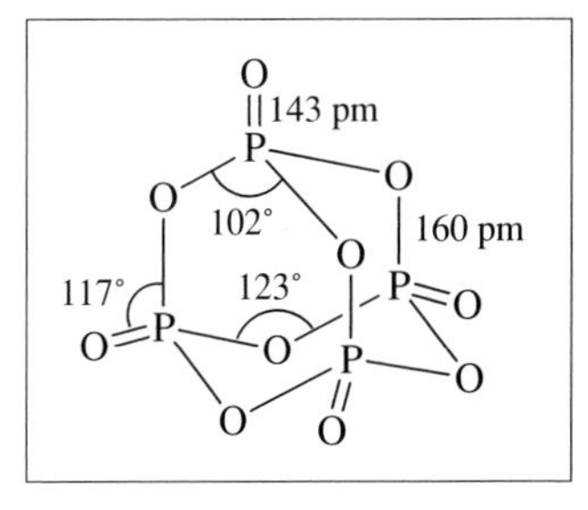

Figure 7.15 The structure of P_4O_{10}

7.3.5 Oxides of the Transition Elements

The well-characterized oxides of the three series of transition elements are shown in Table 7.7. The patterns of oxidation states of the oxides are similar to those of the fluorides of the same elements. The same reasons apply, with the added capacity in some cases for the elements to reach the oxidation state +8. This is mainly because of the removal of the steric factor in the oxides, in which an oxidation state of +8 may be achieved by only four ligand oxygen atoms. These are formally doubly bonded to the metals. The +2 oxides are ionic, as are those of the +3 metals. The oxides of the metals with oxidation states greater than +3 tend to be polymeric, with the highest oxidation state oxides having discrete covalent molecular structures. The lower oxides are basic, the highest ones are acidic, and there are some amphoteric oxides with intermediate oxidation states.

Table 7.7 The oxides of the transition elements

Group									
3	*4*	*5*	*6*	*7*	*8*	*9*	*10*	*11*	*12*
Sc	*Ti*	*V*	*Cr*	*Mn*	*Fe*	*Co*	*Ni*	*Cu*	*Zn*
				Mn_2O_7					
			CrO_3						
		V_2O_5							
	TiO_2	VO_2	CrO_2	MnO_2					
Sc_2O_3	Ti_2O_3	V_2O_3	Cr_2O_3	Mn_2O_3	Fe_2O_3				
	TiO	VO		MnO	FeO	CoO	NiO	CuO	ZnO
								Cu_2O	
Y	*Zr*	*Nb*	*Mo*	*Tc*	*Ru*	*Rh*	*Pd*	*Ag*	*Cd*
					RuO_4				
				Tc_2O_7					
			MoO_3						
		Nb_2O_5							
	ZrO_2	NbO_2	MoO_2	TcO_2	RuO_2	RhO_2			
Y_2O_3						Rh_2O_3			
		NbO					PdO	AgO	CdO
								Ag_2O	
Lu	*Hf*	*Ta*	*W*	*Re*	*Os*	*Ir*	*Pt*	*Au*	*Hg*
					OsO_4				
				Re_2O_7					
			WO_3	ReO_3					
		Ta_2O_5							
	HfO_2		WO_2	ReO_2	OsO_2	IrO_2	PtO_2		
Lu_2O_3						Ir_2O_3		Au_2O_3	
									HgO

Worked Problem 7.9

Q The following table gives the melting points (in °C) for the well-characterized oxides of the first transition series. Interpret these data in terms of possible structures for the compounds.

	Ti	*V*	*Cr*	*Mn*	*Fe*	*Co*	*Ni*	*Cu*	*Zn*
+7				5.9					
+6			196						
+5		690							
+4	1830	1967	300[a]	535[a]					
+3		1970	2266	1080[a]	1565	895[a]			
+2					1369	1795	1984	1326	1975
+1								1235	

[a] Decomposes, loses oxygen.

A The values for Cu^{I} and the +2 oxides of the elements Fe to Zn are consistent with their ionic constitution. This also applies to the +3 oxides, although those of Mn and Co are not proper melting points. The +4 oxides of Ti and V are consistent with their ionic nature, but those of Cr and Mn imply that they are considerably covalent. The higher oxidation state oxides of V, Cr and Mn become less polymeric along the series, with the Mn^{VII} oxide being molecular.

7.3.6 Oxides of the f-Block Elements

Oxides of the Lanthanides

All the lanthanide elements form +3 oxides which are strongly basic. Cerium, Pr and Tb also form formally +4 oxides, but only the Ce oxide approaches 1:2 stoichiometry. The oxides of Pr and Tb are **non-stoichiometric** and approximate to the formulae Pr_6O_{11} and Tb_4O_7, both showing **oxygen deficiencies**. The three "dioxides" have the fluorite (CaF_2) structure with oxide ion vacancies. They are basic in character. Three of the lanthanide elements, Sm, Eu and Yt, form +2 oxides with sodium chloride structures. They are basic.

Oxides of the Actinides

All the actinide elements form +3 oxides similar to those of the lanthanides. The elements which have additional oxidation states in their binary compounds with oxygen are shown in Table 7.8, together with their +3 oxides. The oxides of the actinides show a pattern of oxidation states similar to those exhibited by the fluorides. The oxides of the ear-

Table 7.8 The oxides of the actinides

Ac	*Th*	*Pa*	*U*	*Np*	*Pu*	*Am*	*Cm*	*Bk*	*Cf*
			U_3O_8	Np_3O_8					
		Pa_2O_5	a	a					
	ThO_2	PaO_2	UO_2	NpO_2	PuO_2	AmO_2	CmO_2	BkO_2	CfO_2
Ac_2O_3					Pu_2O_3	Am_2O_3	Cm_2O_3	Bk_2O_3	Cf_2O_3
	ThO	PaO	UO	NpO	PuO	AmO		BkO	

[a] Indicates that there is an almost infinitesimal transition between the compositions of the upper and lower oxides observed in these cases.

lier actinides are more similar to a series of transition elements, but beyond the half-filled stage of the 5f orbitals the series become similar to the oxides of the corresponding lanthanides. The lower oxidation states are basic, the higher ones acidic, with intermediate oxidation states being amphoteric.

Summary of Key Points

1. The periodicities of the main characteristics of the fluorides and the oxides of the elements were described, and the acid/base characters of the oxides were described.

2. Across the periods there is a transition from ionic to covalent compounds, with elements showing their expected valencies/oxidation states, and with some of the p-block elements showing the inert pair effect and/or hypervalency. The transition elements exhibit a wide range of oxidation states in both fluorides and oxides, with the oxides varying from basic to acidic as the oxidation state of the transition element increases.

Problems

7.1. Do the melting points of the fluorides of the third period indicate the type of bonding occurring? The melting points of the fluorides are:

NaF	MgF_2	AlF_3[a]	SiF_4	PF_5	SF_6	ClF_5
1012 °C	1263 °C	1272 °C	–86 °C	–75 °C	–64 °C	–113 °C

[a] AlF_3 sublimes at this temperature to give Al_2F_6, an F-bridged dimer.

7.2. Do the melting points of the s- and p-block (higher valency) fluorides of the fifth period indicate the type of bonding occurring? The melting points of the fluorides are:

RbF	SrF_2	InF_3	SnF_4	SbF_5	TeF_6	IF_7
775 °C	1400 °C	1150 °C	705 °C	7 °C	–35 °C	5 °C

7.3. Comment on the melting point data for the fluorides of rhenium:

RhF_4[a]	RhF_5	RhF_6	RhF_7
>300 °C	48 °C	18.5 °C	48.3 °C

[a] Sublimes.

7.4. Comment on the possible structures and bonding of the following oxides from the melting point data given:

Cs_2O	490 °C	Ta_2O_5	1870 °C
BaO	1920 °C	WO_3	1473 °C
Lu_2O_3	2487 °C	Re_2O_7	296 °C
HfO_2	2900 °C	OsO_8	40 °C

7.5. Comment on the possible structures and bonding of the following fluorides from the melting point data given:

RbF	SrF_2	YF_3	ZrF_4[a]	NbF_5	MoF_6
775 °C	1400 °C	1152 °C	908 °C	78 °C	17 °C

[a] Sublimation temperature.

7.6. BF_3 and AlF_3 have melting points of –127 °C and 1272 °C, respectively. Comment on the difference in bonding that explains the difference.

Further Reading

U. Müller, *Inorganic Structural Chemistry*, 2nd edn., Wiley, New York, 1992. A comprehensive and understandable treatment of inorganic structures.

F. A. Cotton, G. Wilkinson, C. A. Murillo and M. Bochmann, *Advanced Inorganic Chemistry*, 6th edn., Wiley, New York, 1999. A very advanced account of inorganic chemistry, but one in which the detailed descriptions of many inorganic compounds may be found.

Answers to Problems

Chapter 1

1.1. RAM(Cr) = (49.9461 × 0.0435) + (51.9405 × 0.8379) + (52.9407 × 0.095) + (53.9389 × 0.0236) = 51.9959.

1.2. (i) $E = hc/\lambda = 6.6260755 \times 10^{-34}$ J s × 299792458 m s^{-1}/470 × 10^{-9} m = 4.2265×10^{-19} J. (ii) $E = N_A hc/\lambda = 6.0221367 \times 10^{23}$ mol^{-1} × 4.2265×10^{-19} J = 254.5 kJ mol^{-1}.

1.3. $\nu = W/N_A h$; the threshold frequencies for Au, V, Mg and Ba are 1.233 PHz, 1.04 PHz, 0.8846 PHz and 0.6541 PHz, respectively [1 PHz = 1 petahertz = 1×10^{15} Hz].

1.4. The graph (drawn on paper or calculated by spreadsheet) has an intercept on the frequency axis of 0.5562 PHz, which converts to a work function of 221.9 kJ mol^{-1} for the potassium atom. The slope of the line is $N_A h$, and gives a value for Planck's constant of 6.6261×10^{-34} J s.

1.5. Using equation (1.16) with $n_i = 2$ for the Balmer series and $n_j = 3$ for the line of highest frequency (longest wavelength), a value for the Rydberg constant is calculated as: $36/5\lambda = 36/(5 \times 656.3)$ and does not have to be evaluated.
(i) $n_i = 2$ and $n_j = 4$ for the second Balmer line, and the wavelength is obtained by solving the equation:

$$\frac{1}{\lambda} = \frac{36}{5 \times 656.3}\left(\frac{1}{2^2} - \frac{1}{4^2}\right)$$

which gives the wavelength as 486.1 nm.
(ii) $n_i = 3$ and $n_j = 4$ for the first Paschen line and the wavelength is obtained by solving the equation:

$$\frac{1}{\lambda} = \frac{36}{5 \times 656.3}\left(\frac{1}{3^2} - \frac{1}{4^2}\right)$$

which gives the wavelength as 1875 nm.
(iii) $n_i = 1$ and $n_j = 2$ for the first Lyman line and the wavelength is obtained by solving the equation:

$$\frac{1}{\lambda} = \frac{36}{5 \times 656.3}\left(\frac{1}{1^2} - \frac{1}{2^2}\right)$$

which gives the wavelength as 82.3 nm.

Chapter 2

2.1. The degeneracies of the 3p, 4d and 5d atomic orbitals in a hydrogen atom (or any other atom) are 3, 5, and 5, respectively. The degeneracy is given by the value of $2l + 1$ for any value of l.

2.2. d orbitals are associated with values of $l = 2$, and the number of d orbitals is given by $2l + 1$ for any value of l; therefore when $l = 2$, $2l + 1 = 5$, which is the number of permitted d orbitals. If l is less than 2 there cannot be any d orbitals.

2.3. f orbitals are associated with values of $l = 3$ and the number of f orbitals is given by $2l + 1$ for any value of l; therefore when $l = 3$, $2l + 1 = 7$, which is the number of permitted f orbitals. If l is less than 3 there cannot be any f orbitals.

2.4. There are radial nodes in the 3s atomic orbital at values of ρ of 14.196 and 3.804, or values of r of 375.6 pm and 100.6 pm. These values are where the RDF is zero.

Chapter 3

3.1. The ionization energies of the ions He^+, Li^{2+} and Be^{3+} are given by equation (3.11) by putting $n = 1$ (because the $n = \infty$ value is zero energy) and putting the values of Z as 2, 3 and 4, respectively, with $I_H = 1312$ kJ mol^{-1}. This gives 5248, 11808 and 20992 kJ mol^{-1}.

3.2. The titanium atom has the electronic configuration $[Ar]4s^2 3d^2$ because it minimizes the d–d interelectronic repulsion energy. The alternative configuration $3d^4$ would be of higher energy. In the terminology used in the chapter, the two configurations would have energies given by:

$$E(s^2d^2) = 2E_{4s} + 2E_{3d} + J_{ss} + J_{dd} + 4J_{sd} - K_{dd} - 2K_{sd}$$
$$E(d^4) = 4E_{3d} + 6J_{dd} - 6K_{dd}$$

and the six J_{dd} terms in the second equation are sufficient to overcome any stabilization due to the higher negative value of E_{3d}.

3.3. The electronic configurations of the elements of the third period are:

Element	*1s*	*2s*	*2p*	*3s*	*$3p_x$*	*$3p_y$*	*$3p_z$*
Na	⇅	⇅	⇅	↑			
Mg	⇅	⇅	⇅	⇅			
Al	⇅	⇅	⇅	⇅	↑		
Si	⇅	⇅	⇅	⇅	↑	↑	
P	⇅	⇅	⇅	⇅	↑	↑	↑
S	⇅	⇅	⇅	⇅	⇅	↑	↑
Cl	⇅	⇅	⇅	⇅	⇅	⇅	↑
Ar	⇅	⇅	⇅	⇅	⇅	⇅	⇅

3.4. The detailed electronic configurations of the elements V, Cr, Mn and Fe are:

Element	*4s*	*$3d_{xy}$*	*$3d_{xz}$*	*$3d_{yz}$*	*$3d_{x^2-y^2}$*	*$3d_{z^2}$*
V	⇅	↑	↑	↑		
Cr	↑	↑	↑	↑	↑	↑
Mn	⇅	↑	↑	↑	↑	↑
Fe	⇅	⇅	↑	↑	↑	↑

3.5. The seven 4f electrons are distributed according to Hund's rules and occupy the 4f orbitals singly with parallel spins.

3.6. The properties of the elements of the 2nd period and those of the elements of the first transition series are atypical of their "group chemistry", and the properties of the lanthanides are significantly different from those of the actinides because the valence orbitals (2p in the case of the elements of the 2nd period, 3d in the case of the first row of transition elements and the 4f in the case of the lanthanides) are not shielded from the nuclear charge by orbitals of the same spatial arrangements. This is because 1p, 2d and 3f orbitals are forbidden to exist by the quantum rules.

Chapter 4

4.1. Element A shows a large increase of ionization energy between ionizations 4 and 5 and is a member of Group 14; Group 4 elements show a more gradual increase. The large fifth ionization energy is associated with ionization from a lower electron shell. From the high value of the first ionization energy the element is probably the first in the group, carbon. Element B shows a relatively large increase between ionization energies 3 and 4 and is a Group 15 element. The first three ionization energies are those in which p electrons are removed, the fourth and fifth values arising from the ionizations of the two s electrons. Element C shows a very low first ionization energy, followed by a large increase, two smaller increases and then a final larger increase. The element is from Group 1, the first ionization being that of the solitary valence electron, the others a removal of the next lower p electrons facilitated by interelectronic repulsion; the final ionization energy is that from a singly occupied p orbital.

4.2. The sums of the ionic radii for the P^{V}–O and Mn^{VII}–O interactions are 198 and 179 pm, respectively. These values are considerably greater than the observed interatomic distances and mean that the bonding in both ions must be considerably covalent. That the observed interatomic distances are also smaller than the sums of the appropriate covalent radii for single bonds means that the bonds in both ions have bond orders greater than one.

4.3. The first and second ionization energies of the Group 1 elements are, in terms of the electronic changes that occur:

Group 1 element	*First ionization energy*/kJ mol^{-1}	*Second ionization energy*/kJ mol^{-1}
Li	$[He]2s^1 \rightarrow [He]$	$[He] \rightarrow 1s^1$
Na	$[Ne]3s^1 \rightarrow [Ne]$	$[Ne] \rightarrow 2s^22p^5$
K	$[Ar]4s^1 \rightarrow [Ar]$	$[Ar] \rightarrow 3s^23p^5$
Rb	$[Kr]5s^1 \rightarrow [Kr]$	$[Kr] \rightarrow 4s^24p^5$
Cs	$[Xe]6s^1 \rightarrow [Xe]$	$[Xe] \rightarrow 5s^25p^5$

(i) The decrease in the values of the first ionization energies down the group are because of the less effective nuclear charge (because of shielding from completed energy levels) as the atom becomes larger.
(ii) The very high values of the second ionization energies are because the electrons removed originate in the next lower shell, and with Li in particular there is a very large difference because the second electron comes from the 1s orbital, whereas in the other elements the second electron originates in a p orbital.
(iii) The decrease in the values of the second ionization energies down the group are a combination of the two reasons already given. That the large value for Li is due to the large 2s–1s energy gap, the other values are not as high because the electronic change depends upon the $(n + 1)$s to np gap, and this gap is affected by the general decrease in the effectiveness of the nuclear charge.

4.4. Using the supplied data and the given formula the densities (kg m^{-3}) of the Group 2 elements are:

Be	*Mg*	*Ca*	*Sr*	*Ba*
1879	1719	1535	2586	3622

The irregular trend down the group is because of the different ways in which the RAM and r_{ionic} vary. The RAM values increase faster than the ionic radii, and the interactions between the two parameters cause the minimum density to be at calcium [although density is not treated in this book, the densities of the elements do show a periodicity that is worthy of study].

Chapter 5

5.1. Non-metallic elements which exist as infinite three-dimensional arrays require chemical bonds to be broken before the solid can be melted, and van der Waals forces would be expected to play a very small part in their cohesion; they should therefore have high values of m.p. Elements consisting of polymeric sheets have a greater cohesion from van der Waals forces, but chemical bonds still have to be broken to melt the solid. Elements consisting of polymer chains have an even greater input from van der Waals forces, and less from chemical bonding, in causing their physical constants to be smaller than those of the previous categories. Small molecules are held together in the solid by van der Waals forces, and the breaking of these is often sufficient to cause melting and boiling. The diatomic elements are the easiest of those forming molecules to melt and boil; they have relatively small RMM values and the van der Waals forces are small. As examples of these five categories, the melting points (K) of diamond, black phosphorus, selenium, white phosphorus and dinitrogen are 3820, 610, 490, 317 and 63, respectively.

5.2. Taking the oxidation state of the hydrogen atom to be +1, as H is less electronegative than N, the oxidation state of the nitrogen atom is –3.

5.3. In NF_3 the fluorine is the more electronegative atom and takes the oxidation state –1; thus the oxidation state of the nitrogen atom is +3. In both NH_3 and NF_3 the nitrogen is trivalent.

5.4. Boron and carbon form infinite arrays with three and four electrons per atom contributing to the bonding, but nitrogen forms the diatomic dinitrogen molecule which is held in the solid by weak van der Waals forces. Aluminium is a metal and is more easily melted than is boron. Silicon–silicon bonds are weaker than C–C bonds. White phosphorus has P_4 molecules held in the solid by stronger van der Waals forces than the diatomic molecules of N_2.

5.5. The oxidation state of the chloride ion is –1 so that of Pt in $[PtCl_4^{2-}]$ is Pt^x where $x - (4 \times -1) = -2$; therefore $x = +2$; and that of Au in the $[AuCl_4^-]$ is Au^y where $y - (4 \times -1) = -1$; therefore $y = +3$.

Chapter 6

6.1. The phosphorus molecule is P_4, so its atomization enthalpy is 4 × 316.5 = 1266 kJ (mol P_4)$^{-1}$. To transform the element in its standard state to the vapour phase without any bond dissociation requires 58.9 kJ (mol P_4)$^{-1}$. The difference between the two enthalpies represents the amount of enthalpy required to cause the dissociation of the six P–P bonds of the tetrahedral molecule, *i.e.* 1266 – 58.9 = 1207.1 kJ mol^{-1}, so the P–P bonds each require 1207.1/6 = 201 kJ mol^{-1} (to the nearest whole number). The enthalpy of sublimation of the element of 58.9 kJ (mol P_4)$^{-1}$ represents the enthalpy used to overcome the intermolecular forces when phosphorus is melted and evaporated. It represents 5890/1266 = 4.7% of the total enthalpy of atomization.

6.2. The iodine molecule is I_2 so its atomization enthalpy is 2 × 107 = 214 kJ (mol I_2)$^{-1}$. The dissociation of the I–I bond requires 151 kJ mol^{-1}. The difference between the two enthalpies is 214 – 151 = 63 kJ mol^{-1}, which represents the amount of enthalpy required to overcome the intermolecular forces when iodine is melted and evaporated. It represents 6300/214 = 29.4% of the total enthalpy of atomization. Compared to the values calculated for P and S, this is a high percentage and implies some extra cohesion occurring in I_2(s).

6.3. No, but the metallic sheen is a possible indication of some long-range interaction between the constituent diatomic molecules. The I–I distance in the molecules of the solid state is 271.5 pm (longer than in the gas phase, 266.6 pm), with nearest I neighbours at distances of 350 pm and 397 pm in the layer structure that it adopts. The distance between layers is 427 pm. The solid is a poor conductor of electricity, although in pure crystals there is some semiconduction in the directions of the planes in the structure. At very high pressures (*i.e.* above 350 kbar) the solid becomes metallic and shows proper metallic conduction which decreases with increasing temperature.

6.4. If the two sets of data are plotted against the group number it can be seen that there is a very good correlation between them. This is because melting is associated with the loosening of the metallic bonding and much lesser contributions from London forces. At the boiling point, all metallic bonding ceases as gas phase atoms are formed, and the boiling points would be expected to show an even better correlation (look them up in Emsley's book). Plot both m.p. and b.p. data against the values of the enthalpies of atomization to obtain a statistically better idea of the goodness of the correlations.

6.5. The discontinuity from Yb to Lu is because of the changes in electronic configuration when their third electrons are removed. For Yb the change is $[Xe]4f^{14}$ to $[Xe]4f^{13}$, with the loss of $6K$ of exchange energy. For Lu the 4f orbitals remain completely full, and there is no loss of exchange energy. The change in configuration is $[Xe]4f^{14}5d^1$ to $[Xe]4f^{14}$, the ionization from the higher energy 5d orbital providing an added factor in causing the discontinuity.

Chapter 7

7.1. Yes. The values for NaF, MgF_2 and AlF_3 are those expected for ionic compounds. The increase in m.p. from NaF to MgF_2 is expected from the doubly charged Mg^{2+} ion, but the value for AlF_3 shows very little increase and this means that there is a considerable partial covalency present. The sharp fall in melting points in the other fluorides marks the transition to discrete molecular (*i.e.* covalent) forms.

7.2. Yes. The first two values are consistent with the ionic nature of the compounds. The lower values for GaF_3 and SnF_4 indicate the onset of covalency in the polymeric structures. The sharp fall in melting points in the other fluorides marks the transition to discrete molecular (*i.e.* covalent) forms.

7.3. The data are consistent with the transition from covalent polymeric +4 to the discrete higher oxidation state molecular compounds.

7.4. All the metal atoms in the oxides have the [Xe] core electronic structure if they are regarded as fully ionic compounds. The bonding is expected to vary from ionic to covalency as the value of the oxidation state of the metal increases. The first four oxides have melting points which increase with the oxidation state of the central metal atom, and are ionic. The metal ions are surrounded by at least six oxide ions, except for Cs_2O where the coordination number of the Cs^+ ions is only three. The decrease in going to the Ta oxide indicates that there is considerable covalency in the bonding, and the Ta^V can only arrange to have a higher coordination number (six) by forming a polymeric structure. This tendency continues in WO_3 in which the W atoms seek a coordination number of six, the oxide being polymeric. The Re^{VII} oxide melts at a much lower temperature than WO_3, but is still polymeric in the solid to allow the Re atoms to have a higher coordination number. Only in OsO_4 are there sufficient oxide ions/oxygen atoms to allow the central metal atom to have covalent saturation consistent with forming four double bonds, and the compound is truly molecular.

Thus there is a general tendency for compounds to be in the sequence: ionic, covalent polymeric, covalent molecular, as the oxidation state of the central atom increases. Along such a series the melting points (and boiling points) at first increase as the lattice enthalpy of the ionic compounds increase with metal oxidation state. They go through a maximum as the transition from ionic to covalent bonding occurs. The intermediate compounds are covalent polymeric to allow the central metal atoms to gain their octet or 18-electron shares, and only at the end of the sequence is there a steep decrease in the magnitude of the physical constants, as compounds are formed in which the central metal is covalently saturated and which are truly molecular.

7.5. As in Problem 7.4 there is a change from ionic fluorides (Rb and Sr) to the onset of significant covalency in YF_3, more covalency in ZrF_4 which is polymeric, NbF_5 which is a tetrameric oligomer, to the molecular MoF_6.

7.6. BF_3 is a covalent molecule in which the boron atom achieves a share in an octet of valence electrons by accepting a pair from the ligand fluoride atoms. Boron is too electronegative to engage in ionic bonding, but aluminium is sufficiently electropositive to be a metal and its anhydrous fluoride is ionic.

Subject Index